Dynamics of Disaster

Science in Society Series

Series Editor: Steve Rayner
Institute for Science, Innovation and Society, University of Oxford

Animals as Biotechnology
Ethics, Sustainability and Critical Animal Studies
Richard Twine

Business Planning for Turbulent Times
New Methods for Applying Scenarios
Edited by Rafael Ramírez, John W. Selsky and Kees van der Heijden

Debating Climate Change
Pathways through Argument to Agreement
Elizabeth L. Malone

Democratizing Technology
Risk, Responsibility and the Regulation of Chemicals
Anne Chapman

Genomics and Society
Legal, Ethical and Social Dimensions
Edited by George Gaskell and Martin W. Bauer

Influenza and Public Health
Learning from Past Pandemics
Edited by Tamara Giles-Vernick and Susan Craddock, with Jennifer Gunn

Marginalized Reproduction
Ethnicity, Infertility and Reproductive Technologies
Lorraine Culley, Nicky Hudson and Floor van Rooij

Nanotechnology
Risk, Ethics and Law
Edited by Geoffrey Hunt and Michael Mehta

Resolving Messy Policy Problems
Handling Conflict in Environmental, Transport, Health and Ageing Policy
Steven Ney

Rationality and Ritual
Participation and Exclusion in Nuclear Decision-making
Brian Wynne

The Limits to Scarcity
Contesting the Politics of Allocation
Edited by Lyla Mehta

Uncertainty in Policy Making
Values and Evidence in Complex Decisions
Michael Heazle

Unnatural Selection
The Challenges of Engineering Tomorrow's People
Edited by Peter Healey and Steve Rayner

Vaccine Anxieties
Global Science, Child Health and Society
Melissa Leach and James Fairhead

A Web of Prevention
Biological Weapons, Life Sciences and the Governance of Research
Edited by Brian Rappert and Caitrìona McLeish

Dynamics of Disaster

Lessons on Risk, Response and Recovery

Edited by Rachel A. Dowty and Barbara L. Allen
with a foreword by Alan Irwin

earthscan
from Routledge

First published in paperback 2024

First published by Earthscan in the UK and USA in 2011

For a full list of publications please contact:
Earthscan
4 Park Square, Milton Park, Abingdon, Oxon OX14 4RN
605 Third Avenue, New York, NY 10158

Earthscan is an imprint of the Taylor & Francis Group, an informa business

Publisher's Note
The publisher has gone to great lengths to ensure the quality of this reprint but points out that some imperfections in the original copies may be apparent.

A catalogue record for this book is available from the British Library

Library of Congress Cataloging-in-Publication Data

Dowty, Rachel.
Dynamics of disaster : lessons on risk, response, and recovery / Rachel Dowty and Barbara Allen ; with a foreword by Alan Irwin.
p. cm.
Includes bibliographical references and index.
ISBN 978-1-84971-143-2 (hardback : alk. paper) 1. Disaster relief. 2. Disaster victims—Services for. 3. Disaster relief—Economic aspects. 4. Disaster relief—Government policy. 5. Emergency management. I. Allen, Barbara L. II. Title.
HV553.D685 2010
363.34—dc22

2010038148

ISBN: 978-1-84971-143-2 (hbk)
ISBN: 978-1-03-292578-3 (pbk)
ISBN: 978-1-315-06719-3 (ebk)

DOI: 10.4324/9781315067193

Typeset by Composition and Design Services
Cover design by Susanne Harris

Contents

Part III International Disasters and Katrina Comparisons

List of Figures, Tables and Boxes

Figures

Tables

Boxes

List of Acronyms and Abbreviations

BNOB	Bring New Orleans Back
BQP	Bitterfelder Qualifizierungs- und Projektierungsgesellschaft (Bitterfeld Qualification and Project Management Association)
CBD	central business district
CHAT	New Orleans Wiki Citizens' Road Home Action Team
CHNO	Covenant House New Orleans
DEQ	State of Louisiana Department of Environmental Quality
DHS	Department of Homeland Security
DKKV	German Committee for Disaster Reduction
DRK	German Red Cross Society
EPA	Environmental Protection Agency (US)
FCO	Federal Coordinating Officer
FEMA	Federal Emergency Management Agency
GIS	geographic information system
GPS	global positioning system
HANO	Housing Authority of New Orleans
HDLC	Historic District Landmarks Commission
HUD	Department of Housing and Urban Development
IPCC	Intergovernmental Panel on Climate Change
ITE	intelligent trial and error
LAF	Landesanstalt für Altlastenfreistellung (State Authority for Exemption from Residual Pollution Liability)
LTTE	Liberation Tigers of Tamil Eelam
MDSE	Mitteldeutsche Sanierungs- und Entsorgungsgesellschaft (Central German Association for Remediation and Disposal)
MR-GO	Mississippi River Gulf Outlet
NASA	National Aeronautic and Space Administration
NGO	non-governmental organization
NNI	National Nanotechnology Initiative
NOAA	National Oceanic and Atmospheric Administration
NPN	Neighbourhoods Planning Network
NPS	National Park Service
NPSS	new political sociology of science

NSF	National Science Foundation
PFO	Principal Federal Official
PRC	Preservation Resource Center
PUS	public use/understanding of science
QTRP	Qatar Tremé Renewal Project
RHP	Road Home Programme
RT	Rebuilding Together
SBA	Small Business Administration
SGER	Small Grants for Exploratory Research
SHPO	State Historic Preservation Office
STF	Special Task Force
STS	Science and Technology Studies
UNOP	Unified New Orleans Planning

Contributors

Barbara L. Allen is Associate Professor and Director of the Science and Technology Studies graduate programme at Virginia Tech's Washington DC area campus.

Roberto E. Barrios is Assistant Professor in the Department of Anthropology at Southern Illinois University, Carbondale.

Colin E. Beech is a Policy Researcher in the Office of Research and Sponsored Programs at the University of New Orleans, Louisiana.

Alena Bleicher is a PhD student and junior research scientist in the Department of Urban and Environmental Sociology at Helmholtz Centre for Environmental Research in Leipzig, Germany.

Gary Bowden is Associate Professor in the Department of Sociology at the University of New Brunswick, Fredericton, New Brunswick, Canada.

Emilie Brévière is the Executive Officer of SOLAS (Surface Ocean–Lower Atmosphere Study) at the School of Environmental Sciences, University of East Anglia, UK.

Vivian Y. Choi is a doctoral candidate in the Department of Anthropology at the University of California, Davis.

Rachel A. Dowty is Assistant Professor-Research in the Department of Geography and Anthropology and Co-Director of the Disaster Science and Management (DSM) Program at Louisiana State University.

Scott Frickel is Associate Professor and Co-Director of Graduate Studies in the Department of Sociology at Washington State University and Adjunct Faculty in the Department of International Health and Development at Tulane University School of Public Health & Tropical Medicine.

Melanie Gall is Assistant Professor-Research in the Department of Geography and Anthropology and teaches in the Disaster Science and Management programme at Louisiana State University.

Matthias Gross is a tenured Senior Researcher in the Department of Urban and Environmental Sociology at Helmholtz Centre for Environmental Research in Leipzig, Germany.

Kathryn Henderson is Associate Professor in Sociology at Texas A&M University, College Station, Texas.

Alan Irwin is Professor in the Department of Organisation and Dean of Research at Copenhagen Business School.

Isabelle Thomas Maret is Associate Professor in the Institut d'urbanisme at the Université of Montréal.

Peter J. May is the Donald R. Matthews Distinguished Professor American Politics, Department of Political Science of the University of Washington, Seattle.

M. Bess Vincent is a doctoral candidate in the Department of Sociology at Tulane University.

William (Al) Wallace is Yamada Corporation Professor of the Department of Industrial and Systems Engineering, with joint appointments in the Civil and Environmental Engineering, and Cognitive Science Departments, and is Director of Rensselaer's Center for Infrastructure and Transportation Studies at Rensselaer Polytechnic Institute, New York.

Juergen Weichselgartner is the Senior Science Coordinator of LOICZ (Land-Ocean Interactions in the Coastal Zone) at the Institute for Coastal Research, Helmholtz-Zentrum Geesthacht, Germany.

Edward J. Woodhouse is Professor in the Department of Science and Technology Studies at Rensselaer Polytechnic Institute, New York.

Acknowledgements

The editors wish to thank the many people whose work made this volume possible. Joseph Powell at Louisiana State University was the first to suggest that the Society for the Social Studies of Science (4S) conference panel organized by Rachel Dowty be turned into a co-edited volume with Barbara Allen. Prof. David Hess of Rensselaer Polytechnic Institute organized the 2006 4S conference in Montréal where the conference panel took place. We thank the original conference panelists (Edward Woodhouse, Kathryn Henderson, Roberto Barrios and Gary Bowden) for participating in the panel and continuing to develop their respective papers for this book. We take our hats off to Scott Frickel, M. Bess Vincent, Isabelle Maret, Juergen Weichselgartner, Emilie Brévière, Melanie Gall, Vivian Choi, Alena Bleicher and Matthias Gross for joining our efforts and contributing their scholarship. Alan Irwin offered his knowledge and insight in the wonderful Foreword he contributed to this volume.

Rachel Dowty wishes to thank the Department of Science and Technology Studies and Professor William 'Al' Wallace in the Department of Decision Sciences and Engineering Systems at Rensselaer Polytechnic Institute for encouraging her interdisciplinary work in disaster sciences. She would also like to thank Professor Peter May at the University of Washington for his interest and thoughtful feedback on this work. Also, Louisiana State University's College of Arts and Sciences, the Department of Geography and Anthropology, and the Disaster Science and Management (DSM) programme has offered ongoing support of this and other related projects. Special thanks go to Professor Patrick Hesp at LSU for his continuing support of the DSM programme.

Barbara Allen wishes to thank the National Science Foundation (grant no. SES-0821353) for enabling her to pursue interdisciplinary research into successful post-disaster rebuilding strategies after Hurricane Katrina from a Science and Technology Studies perspective. Additionally, the Liguria Study Center of the Bogliasco Foundation provided time and space for the early development of this book. Virginia Tech provided her with valuable research leave and the University of Michigan, the University of Padova and Columbia University have provided her with opportunities to lecture and receive engaged responses to the work. She would especially like to thank PhD student, Sumitra Nair, who helped to carefully edit the final manuscript.

We are both very grateful to Earthscan's Alison Kuznets for her enthusiasm, helpful suggestions and guidance throughout the development and publication process.

Foreword

Alan Irwin

How are we to think about disasters? The whole physical and societal impact of an event such as Hurricane Katrina seems to stem from its inherently shocking nature, and its generally 'out of the ordinary' character. Viewed alongside the depressing toll of human deaths and injuries, televised images of apparent chaos, social breakdown and obvious misery proved deeply disturbing to national and international viewers. For those with fond recollections, or simply mental associations, of New Orleans as a place of distinctive (perhaps subversive) charm and lively culture, nature's sudden impact was hard to comprehend. Who would have thought that a global superpower could so closely resemble a developing nation? Reflecting the international concern provoked by the pictures of stranded, abandoned and inadequately housed residents, inconsistent relief efforts and manifest poverty, offers of foreign aid began to pour in (prompting, as Allen and Maret record below, a terse response from the Bush administration: 'Thanks, but no thanks.'). Other events such as the 2002 flood in the Elbe Region of Germany, the Indian Ocean tsunami and the 2010 Haitian earthquake reinforce this sense of disasters as operating in a realm 'beyond the everyday': a view only strengthened by the vivid accounts provided by the global media. For that reason also, we struggle to think of what might be termed 'slow disasters' (the health impacts of poverty, avoidable deaths from infectious diseases, the consequences of long-term environmental pollution) as fitting into that category at all, since (apparently) they lack immediacy, high drama and the shock of the unexpected.

As this book richly explores, there are many reasons why the view of disasters as 'natural accidents' cannot suffice. All the chapters here bring home the basic point that disasters do not operate in a realm beyond the ordinary but instead resonate with and reflect our ways of life and the conditions we create: in that sense, disasters are indeed 'normal'. As Mary Douglas might have put this, disasters cannot be separated from the societies in which they occur. Social structures and ways of living in turn serve to define what counts as a disaster and even the form that it takes. Thus, Allen and Maret relate that Hurricane Katrina has become 'iconic' among environmental justice scholars and activists precisely because of its external uncovering of deeply rooted racial disparities in the US – disparities which appear to have contributed to the devastating impact but which also provided the context for subsequent recovery efforts. The point of course is that the hurricane did not create these inequalities, nor would their existence be a surprise to most residents in the affected areas. Far from standing alone, the disaster formed one element within the locally constituted cultural–natural mix. Woodhouse makes this point very directly when he presents disasters as 'extreme versions of everyday life'. Seen from that perspective, the production of disasters should be seen as 'not all that special' since they represent not the sudden appearance of one large problem but rather the steady accumulation of lots of small, everyday concerns. Choi makes a related point in eloquent fashion: 'In Sri Lanka, as in New Orleans after Hurricane Katrina, when the dark and dangerous waters of the tsunami receded, the disaster laid bare some of the most embedded of social issues' (Chapter 10, this volume).

What 'embedded issues' can we therefore identify across these cases of 'risk, response and recovery'? Of course, one crucial merit of cross-cultural (and cross-disaster) comparisons is that they can alert us to what Weichselgartner and Brévière describe as 'context-sensitive, and historically and socially contingent knowledge'. The social and environmental conditions of New Orleans are not the same as those experienced by elderly Parisians during the 2003 heatwave. The recovery effort after the Elbe floods was very different to that which took place following Hurricane Katrina. In one case, we find expressions of community anger coupled with charges of institutional irresponsibility and dereliction of duty. In the other, disaster response appeared to represent an enactment of national solidarity based on the notion that no one should be worse off as a consequence. Beneath such crucial differences and the inevitable messiness of human experience, however, common themes emerge – only some of which can be hesitantly captured here.

As expressed by a number of the authors in this volume, thinking about disasters also means thinking about *the limits of knowing and the assumptions built into expert systems*. For Frickel and Vincent, this suggests the need to understand the shifting constitution of knowledge gaps and, especially, the manner in which 'epistemic efficiency' can lead to a focus on what is already known to the neglect of what is not. Bleicher and Gross frame the embedded issue here in terms of the requirement to act 'in the face of ignorance'. As they put this, the remediation of contaminated land in eastern Germany necessitated an open acknowledgement that ignorance could not be avoided. It was (as metal containers and tar deposits turned up almost at random) a fundamental characteristic of the whole operation rather than a matter for subsequent retribution and blame. Such an approach contrasts sharply with the tendency identified by Weichselgartner and Brévière for traditional risk management to filter out (or gloss over) areas of ignorance and instead to offer a reductionist (and therefore unnecessarily constrained) framework for understanding and control. As these authors express the basic point: 'hazard protection is a socially isolated activity and disaster response an event-focused reaction'

(Chapter 8). One classic problem here, as Brian Wynne and others have pointed out, is that risk management and mitigation planning can draw upon unacknowledged and unchallenged assumptions about the social world, effectively imposing a technicist framework upon more complex webs of action and meaning. Barrios makes this point well in his study of post-Katrina recovery planning: critical differences emerged between the ways in which some residents and planning experts constructed the fundamental notion of a 'neighbourhood' – with important consequences for recovery operations but also for the fundamental understanding of what actually took place during the disaster.

These socio-epistemological observations relate closely to the powerful sense of *place, identity and social vulnerability* that emerges in many of these studies. Henderson approaches this issue through her experience of 'mind maps', and especially of the relationship between situated, contextualized understandings of an area and those provided by 'the Cartesian planes of commercial maps and GPS coordinates'. As she puts this, local understandings of space and of lived boundaries are 'raced and gendered' in complex ways. From this perspective, tacit, visual and situated knowledges tie together closely with a sense of identity: a point which is as true for planning and recovery professionals as it is for local residents. Bowden explores the aftermath of Katrina by asking why (in contrast, for example, to the Elbe case) this disaster led some citizens of New Orleans to play up, rather than minimize, social differences. The answer would appear to lie in differential victim identities which, at least in some cases, connected powerfully with historical legacies of oppression. For Bowden, 'the shared sense of community in New Orleans prior to Katrina was a thin veneer painted over fundamental divisions' (Chapter 3). The point, as in other chapters, is that the experience of disjuncture and dislocation following a disaster can link closely to these multiple senses of the community/neighbourhood. Crucially, and very practically, Gall emphasizes that place and identity can be bound up with levels of social vulnerability and coping capacity. In the cases of both Katrina and the European heatwave, death came most to the elderly, fragile and socially marginalized. In Paris, only 30 per cent of domestic victims were in regular contact with their families.

A third, and again interlinked, theme in these chapters is that of *organizational culture and response*. Thus, Dowty and her colleagues explore in some depth the reasons why there was a considerable variation in institutional performance following Hurricane Katrina, noting especially the cultural biases which led different organizations to see the same disaster situation quite differently. Other chapters echo local criticisms of the post-Katrina response, reinforcing a sense that residents and authorities often held quite incompatible understandings of what was required. Allen and Maret explore both successes and failures within the rebuilding process, concluding that trust (based especially on past interactions and reputation) is a vital element in the relationship between citizens and organizations: when government agencies distrust their own citizens, it becomes extremely hard to build successful cooperation. For these two authors, money alone is not enough in rebuilding neighbourhoods. Trustworthy organizations and democratic politics are an essential part of the mix. As if to reinforce this point, Choi provocatively argues that the management of disaster in Sri Lanka articulates closely with the political and social landscape of war in that country. Presented in this way, both natural disasters and terrorism appear as external, inevitable and uncontrollable occurrences which need to be actively managed by the state. It follows also that events such as the tsunami can engender modes of control that subsequently become applied to other 'non-disaster' contexts. Disaster response can thereby constitute an organizational route to re-ordering (or at least reinforcing) the nation, as apparently more positive cases such as the response to the Elbe flood also seem to suggest.

There are several other themes that could be developed here, including the view of disasters as resulting from the interaction of natural, technological and social systems, the recurrent motif in the following chapters of '(non) learning from experience', and the fundamental question of dealing with socio-technical uncertainty. The editors themselves highlight certain significant elements in the dynamics of disaster, including questions of accountability and responsibility, and the relationship between cosmopolitan ways of acting in the world and cultural specificities. Very ambitiously, Dowty and Allen seek to draw together Science and Technology Studies with what is termed the 'new' political sociology of science alongside a whole range of academic contributions not least from disaster studies but also from social anthropology, sociology, geography and environmental studies (among others). Certainly, there is ample evidence in this book to support the general claim that disasters offer fertile, challenging and evocative territory for a broad range of social scientists.

My fourth and final theme therefore represents an attempt at least to draw attention to (if not exactly capture) these cross-cutting intellectual currents, attempted reflexivities, empirical explorations and calls for accountability. When our understandings of disasters are viewed as operating on the *borderlands of interdisciplinarity*, a number of intellectual and practical challenges come into focus. Drawing on Callon's (1998) distinction between 'hot' and 'cold' to describe the ways in which knowledge is employed within particular contexts, Marilyn Strathern has characterized 'hot' situations as arising from the unpredictable interaction of diverse factors – as in the turmoil of the UK's BSE crisis. The 'heat' in a case such as this stems from the 'mix of quite different orders of knowledge, including that drawn from non-specialists' (Strathern, 2004, p3).

The point here is that 'hot' situations are becoming more common as the boundaries of discipline and skill are crossed (and partly redefined) by recurrent controversies. At the same time (and as the aftermath of Hurricane Katrina nicely demonstrates), 'cooling down' such situations is far from easy since consensus is often hard to find: 'The result is interlocking, scale-crossing complexity. Social scientists are as much caught up in this as anyone' (Strathern, 2004, p4). For the social scientists in question, this means that 'no one person, no one discipline, could grasp all the dimensions' (Strathern, 2004, p4). Instead, and applying Strathern's general argument to the Katrina case, social scientists must engage with activists and community organizations, planners and developers, residents and managers, representatives of different governmental organizations, scientists of many descriptions, and indeed, other sorts of social scientists. Perhaps inevitably also, social scientists become bound up in anticipations of different kinds, recruited (generally willingly) into a state of readiness for the next disaster or crisis. In such circumstances, and as Strathern notes, internal debates between disciplinary experts can be looked upon as a distraction 'and at worst as disciplinary in-fighting or self-indulgence at the expense of human need' (Strathern, 2004, p5). Called upon to represent the 'societal' view among groups of diverse specialists, the social scientist can find herself stranded and vulnerable. As Strathern pointedly expresses it: 'In such circumstances, social scientists might want to think rather carefully about ... just how they are to speak for "society"' (Strathern, 2004, p5).

Put in such terms, the task of what one might term 'engaged but critical social science' can indeed appear daunting. Just like the urban landscape in the aftermath of Hurricane Katrina, the interdisciplinary borderlands are sometimes difficult, uncharted and confusing places, not helped by the fact that it may not even be clear which borders are relevant or what it is that is being separated: is it rich and poor, science and non-science, society and nature, institutional control and chaos, crisis and normality, or some other distinction based on race, gender, organizational perspective or cultural belief? Even the boundaries around disaster are open to challenge: does Chernobyl or Three Mile Island represent a disaster? Does 9/11 or long-term environmental degradation? Meanwhile, and as a number of contributors note, non-certified specialists and lay observers may have much to contribute so that social scientists (and others) cannot claim the monopoly on knowledge and understanding.

Whilst, as Strathern and others have helpfully explored, this situation is decidedly problematic, the chapters that follow also give us some reasons for optimism. When researchers and research participants begin to share problems, new opportunities for action and new intellectual insights can arise. When social scientists maintain critical perspective but also seek to interpret organizational practices in an open fashion, new forms of co-created understanding can develop. When disciplinary boundaries are valued (not least in terms of accounting for knowledge practices) but also serve as a stimulus to new forms of interdisciplinary endeavour, the social scientific imagination can be enhanced and extended. The borderlands can be uncomfortable, bewildering and threatening. They can also be places of regeneration, reconstruction and new possibility.

References

Callon, M. (1998) 'An essay on framing and overflowing: Economic externalities revisited by sociology', in M. Callon (ed) *The Laws of the Markets*, Blackwell Publishers/ The Sociological Review, Oxford

Strathern, M. (2004) *Commons and Borderlands*, Sean Kingston Publishing, Oxford

Introduction

Rachel A. Dowty and Barbara L. Allen

A dynamic approach to disasters requires consideration of multiple voices and perspectives, both human and non-human factors, and interactions not easily given a starting point at the individual or group level. Different ways of life and approaches to knowledge are disrupted by disasters, and daily problems do not disappear but are frequently intensified. Thus, poverty-stricken city residents are still poverty-stricken (or more so than before), governmental inefficiencies and bureaucratic structures still remain (or more so than before), and polluted areas are still polluted (or more so than before). Too frequently, however, disaster responses revolve around very few ways of knowing and ways of living, marginalizing already marginalized voices until they fade into a fuzzy and generalized impression of chaos or mayhem or tragedy. Purveyors of expertise may lose sight of marginalized community voices, especially when they are not personally hit by the disaster. Yet questions loom regarding what constitutes sufficient participation to combat marginalization, as well as what constitutes expertise and policy-relevant knowledge.

People affect each other as well as affecting overall circumstances arising from disasters. It is one thing, for example, to say that New Orleans is a risky place to live, but to say that the people who were stranded by the post-Hurricane Katrina floods were at fault for the situations they found themselves in suggests something more. Who is 'in charge' and why? What mechanisms of accountability do different groups assume are in place, and what differences can be found within those groups regarding such assumptions? Perhaps more importantly, how is accountability for taking risks assessed by different people relative to those same peoples' assessments of disaster response and recovery? Additionally, similar questions can be posed during the process of repatriation and recovery. Who is responsible for managing or enabling reconstruction post-disaster? How are cosmopolitan techniques and technologies coordinated or mediated by local building knowledge and cultural specificities on the ground?

Disaster researchers frequently approach such questions using the concept of vulnerability: which populations are in more vulnerable positions environmentally (Burby, 1998), politically (Boin et al, 2007), socially (Tierney et al,

2001; Perry and Quarantelli, 2005) and culturally (Phillips 1993; Hoffman and Oliver-Smith, 2002). Such vulnerabilities, however, are not mutually exclusive (Rubin, 2007). For example, many people would rather live in less environmentally vulnerable locations (environmental vulnerability), but cannot afford to move (social vulnerability). Some who have political clout are culturally ignorant of the needs of those they serve, and many actions they take in terms of preparedness and mitigation are consequently ineffective and result in broader response and recovery needs.

One of the reasons behind this volume is to bring some new theoretical and methodological approaches to the social and cultural aspects of disaster research from the interdisciplinary field of Science and Technology Studies (STS). These scholars have been at the forefront of asking questions about the politics of technical and scientific knowledge formation and the construction of expertise. While there are numerous STS and allied analytical lenses represented in this volume, we highlight several key ones as particular examples of how they might contribute to disaster studies.

One helpful framing tool from STS is called the 'public use of and understanding of science and technology.' This analytic lens is well developed in the field and has numerous components, but the particular approach of interest to us is understanding how the public and civic cultures shape the formation of risk, scientific knowledge, and technological know-how. Public use/understanding of science (PUS) (Irwin, 1995; Irwin and Wynne, 1996) has been a broad topic of study in the field of Science and Technology Studies (STS) (Hess, 1997; Bauchspies et al, 2005; Kleinman, 2005). It is one of the mechanisms by which citizens participate in the making of technoscience and the shaping of the regulations and policies that govern it. PUS is both the engagement of the lay public in science and technology issues as well as the participation of local or indigenous 'experts' in these issues, often in the heat of controversy or distress (Sclove, 1995; Martin, 1996, 1999; Fischer, 2000). Sometimes the applied local knowledge and practices serve to shift or hybridize cosmopolitan practices making them better suited to on-the-ground problem solving (Eglash et al, 2004). Alan Irwin has conceptualized PUS in a framework of 'citizen science' (1995), a remaking and hybridization of science in terms of people's everyday lives. His questions concerned the forms of scientific/technical knowledge and expertise that were appropriate to the situation and what relationship they had to other forms of knowledge and understanding.

Another STS analytical frame is the 'new' political sociology of science (NPSS) that emerged in the rethinking of the mainstream literature in the politics of science and knowledge (Frickel and Moore, 2006). One of the motivations for the re-examination of this sub-field was the tremendous growth in diversity of lay-people's participation in socio-technical issues, from advocacy to policy making. According to Frickel and Moore (2006, p5), 'We now find a vastly richer array of interactions among scientists, citizens, government, and the private sector than existed in earlier decades.' The constitution of knowledge involves a variety of participants including institutions, experts and citizens – each bringing a set of practices, cultures and biases to the process.

Thus while PUS examines the production of citizen-inflected knowledge, NPSS is focused on institutional knowledge production, particularly such issues as rules and regulations, the dynamics of organizations and networks, and the distribution of resources. This institutional knowledge production is hybrid in nature and can include legal, political and civil society organizations (Kleinman, 2003; Hess, 1997). NPSS, while sensitive to the 'thick descriptions' of the ethnographic tradition in STS, expands the research questions to ask: Why does technoscience and allied knowledge claims work better for some than others? It seeks to find causal explanations for both winners and losers connecting case studies to broader cultural, political and institutional dynamics, making visible the politics of institutional arrangements as well as positive (and negative) policy fulcrums for change.

The authors in this volume analyse social, cultural and political aspects of risk, response and recovery post-disaster in numerous ways that engage, either explicitly or implicitly, some of the analytical lenses we have described above. The sections and chapters involve numerous perspectives on what constitutes rational, equitable and/or 'workable' planning, decision-making and rebuilding/reclamation strategies. This book connects disaster research to STS as well as to environmental studies, sociology and anthropology to approach the complex interactions surrounding natural and man-made disasters. Hurricane Katrina provides a thematic focus for the first two sections because of the general consensus from disaster professionals, victims and spectators that situations arising from the storm raised questions about disaster planning, response and recovery from many interacting perspectives.

Part I of the volume, 'Environmental, Cultural and Political Concerns', is a collection of chapters that could be described as 'voices from above'. In particular, those people whose voices are given authority through legitimization by governmental or other institutions. With regards to Hurricane Katrina, voices from above are those that were responsible for organizing response and recovery activities. 'Relocation, Rebuilding and Recovery Concerns' (Part II) on the other hand, situates and analyses the voices of local citizens, those who were recipients of immediate disaster response and later rebuilding planning and action strategies. Despite these divisions, some voices from above were also victims of the storm and recipients of aid, such that distinctions between voices from above and below must be considered carefully with respect to interactions and fuzzy categorical boundaries that separate people during disaster response and recovery operations. Comparing different perceptions of what constitutes rational decision-making requires examination of Katrina relative to other disaster cases under different orders of authoritative categories (Part III – 'International Disasters and Katrina Comparisons'). The authors in this volume address voices from above, voices from below and voices from outside the situations surrounding Hurricane Katrina to better understand the dynamics of disaster in an international and comparative context.

In Chapter 1, Scott Frickel and M. Bess Vincent offer a definition and characterization of knowledge gaps in the construction of organizational assessments of environmental risk and response to Hurricane Katrina in the

New Orleans area. This involved examination of data generated by the US Environmental Protection Agency's (EPA) Region 6 Katrina and Rita Hurricane Response Project. It identifies sampling strategies that EPA developed to characterize environmental risk in Orleans Parish, Louisiana; these are presented in four, roughly chronological stages. Analysis of that process focuses on the institutional logic that guided decision-making at EPA and which generated a series of knowledge gaps with distinct socio-spatial dimensions. These spatial knowledge gaps demarcate areas of the city where no knowledge was ever produced. Using this information, three moments of contested knowledge surrounding EPA's assessment work are identified to illustrate the (often latent) interrelation of knowledge gaps and contentious politics. The broader political significance of examining regulatory knowledge gaps lies in (1) the capacity to illuminate tensions and contradictions in the institutional logics that govern knowledge production, and (2) issues arising when the same organizations making decisions about what knowledge is needed make decisions about whether and how that knowledge is used.

Rachel Dowty, Peter May, William Wallace and Colin Beech develop social anthropological methods to derive insights into the blame patterns, crisis responses and management emphases of various organizations to Hurricane Katrina in Chapter 2. They also examine ways that leadership from these organizations justified their responses in the face of critical assessments. Analyses of Baton Rouge Emergency Operation Center teleconference transcripts provide a basis for classifying 'cultural biases' of several responding organizations. Results show that the Federal Emergency Management Agency's efforts became fatalistic in coping with overwhelming demands and anticipating criticisms. The US Coast Guard's response reflects the hierarchical notion of devolved authority and sense of purpose in pulling together. Various White House officials reflect an individualistic cultural bias in blaming incompetent bureaucrats ('Brownie') for a poor job. The emergent neighbourhood, church and other volunteer groups reflect an egalitarian bias in feeling that they were the real saviours at the front lines of the rescue effort and in blaming government officials for failing to act. This chapter demonstrates the utility of using an organizational culture lens to examine disaster response, while adding methodological rigour to the cultural understanding of performance shortfalls of public organizations.

Gary Bowden begins Chapter 3 by outlining several tightly interconnected events that would unfold in the city of New Orleans over the few days immediately following Hurricane Katrina's landfall: (1) the devastation from the hurricane itself; (2) the flooding of much of the city resulting from the breeching of various levees by the hurricane's storm surge; and (3) the descent towards a Lord of the Flies, fend for yourself, anarchy (at least for a brief period of time in substantial portions of the city). One of the most consistent findings in sociological studies of disaster is the socially unifying effect of natural disasters. Even after adjusting for the exaggerated media depiction, the fact remains that the resulting social response was highly atypical. How do we account for this fact? The proposed paper answers this question by viewing disasters not as events but, rather, as the actualization of system vulnerabilities. Bowden uses

Perrow's concept of a system accident to highlight the unexpected interaction between the three systems: natural (climate, weather), technological (flood control, transportation and communication) and social. Particular attention is paid to the historic roles of race, poverty and the police in social control in New Orleans and how the flood disturbed normal expectations surrounding patterns of social interaction and led to the emergence of anomic conditions.

In Chapter 4, Edward Woodhouse argues that many of the same processes and institutions are at work to create vulnerable populations, design the built environment carelessly with respect to public values, place barriers in the way of preventive action, and make it difficult for experts to contribute to improved outcomes. Was the Hurricane Katrina disaster an aberration, or did it emerge from decision-making processes similar to those governing other public outcomes? Is it more reasonable to expect post-disaster analyses to lead to systematic learning and improved policy, or not to change very much? Most generally, what can be learned about appropriate expertise and usable knowledge from the Katrina experience? He discusses implications for interdisciplinary, problem-focused research and community service by scientists, engineers and other experts.

Based on interviews and mapping exercises with Hurricane Katrina evacuees, Kathryn Henderson's Chapter 5 employs sketched, Google and overdrawn maps of origin neighbourhoods in comparison with similar format maps of neighbourhoods of relocation to examine the situatedness of local spatial knowledge in daily interaction, its disruption and its rebuilding. Findings are examined in light of Turnbull's analysis of ethno mapping techniques. The research explores the challenges of translating mental maps and visual knowledge onto paper when researcher and participant may not share visual cultures or may use and understand the same visual representations differently. It also explores how everyday spatial perception and real-life lived boundaries are raced and gendered by restrictions of navigation. Techniques are map-elicitation, photo-elicitation, in-depth interviews and participant observation.

Roberto Barrios uses an ethnographic approach to examine experts' attempts to include marginalized voices for recovery planning in Chapter 6. Media representations of the aftermath of Hurricane Katrina in New Orleans called attention to the fundamental inequities that underlie the socio-economic structure of North American cities. Consequently, local government officials, development non-profit organizations and national philanthropic agencies have heralded recovery planning in New Orleans as a 'ground up process' through which the voices of city residents will become the mandate that guides the reconstruction. Analysis of planning documents and ethnographic observations during the drafting process of the United New Orleans Plan provides examination of methodological and political complications of producing such a mandate. The ways expert knowledge has historically been prioritized as an objective means of determining the needs of a population is addressed, relative to the arrangements necessary between people and resources to produce societal well-being. Anthropological writings on the subject have demonstrated how such expert knowledge has been articulated through vernaculars that enact hierarchies of class and race. This chapter explains how architects and planners produce the 'voices of city residents' through so-called participatory planning. It also explains how superficial labels of participatory democracy perpetuate repressive cultural politics that inhabit the making of consensus in recovery planning after Hurricane Katrina.

In Chapter 7, Barbara Allen and Isabelle Maret ask questions about what strategies, alliances, networks 'work best' (and least) to rebuild and repatriate a poor, predominantly African-American community after a disaster. Using the Tremé neighbourhood as a case study they situate their analysis with careful attention to the racial/historical context and spatial practices of a community that, over time, had come to be definitive of 'authentic' Creole culture in New Orleans. Interviews with people from governmental and non-governmental organizations (NGOs), philanthropic groups and community organizations helped identify what they did, how they continue, defining 'success' from a multiplicity of perspectives. These authors identify the institutional dynamics and regulatory/procedural elements that have led to positive rebuilding outcomes, particularly those with an equitable, transparent and more participatory process. Community members were asked the same questions, and comparisons show how well NGO perceptions match citizen's realities towards understanding 'what works' in efforts towards neighbourhood repatriation and recovery.

Beginning the final section of the book on disaster response and management in a comparative and international context is Juergen Weichselgartner and Emilie Brévière's Chapter 8, that gives historical, spatial and political contexts to the 2002 floods in Germany's Elbe river basin. They document events that led up to the flood as historically situated in the way that the region responds to disasters, recent advents in insurance and disaster response, and planning and flood protection activities. When viewed alongside the post-Hurricane Katrina floods in New Orleans, a contrast emerges between the German region's solidarity and the United States' response to flood disasters.

In Chapter 9, Melanie Gall compares Hurricane Katrina and the 2003 European heatwave with regards to social vulnerability. She pays particular attention to what public health experts refer to as 'excess mortality' in the cities of New Orleans and Paris. Vulnerabilities may look different but have different outcomes when viewed from individual, community or organizational levels. Gall examines these two disasters from each of these levels and suggests how to approach issues of social vulnerability to natural disasters that become unnatural in their scope.

Vivian Choi uses Sri Lanka's approaches to managing natural disasters and their approaches for managing war and terrorism to show how one approach reflects another. After the 2004 Boxing Day tsunami, Sri Lanka formed a parliamentary committee to assess natural disaster preparedness and began technological initiatives to improve warning systems and evacuation plans. In Chapter 10, Choi outlines the problems with such initiatives while showing that ongoing political animosities in Sri Lanka cannot be easily separated from natural disaster preparedness and response efforts. When compared to the politics and initial reactions to Hurricane Katrina following the levee breaches in New Orleans, similarities in responses and culturally embedded ideas of risk and preparedness can be seen.

Alena Bleicher and Matthias Gross address 'creeping' disasters as opposed to the more punctuated events addressed in other chapters. In Chapter 11, they discuss disasters that result from polluted former industrial sites in Eastern Germany, approaching risk and uncertainty in terms of 'different shadings of ignorance'. As such, they document decision-making processes regarding environmental management that reflect the sort of knowledge gaps discussed by Frickel and Vincent in Chapter 1, but with an eye to long-term recovery more than to initial response.

Climate change, persistent poverty, increasing population and hazard-producing production as well as general pressure on natural resources are some of the factors behind the increasing likelihood that the world will see more human and natural disasters in the coming years. However, evidenced by the damage done by Hurricane Katrina and comparative disasters in this volume, response and rebuilding can be plagued by a wide range of difficulties. It is our hope that the confluence of STS and other socio-cultural approaches to disaster studies offers some conceptual and research tools that may make it possible to improve post-disaster efforts in the future.

References

Bauchspies, W., Croissant, J. and Restivo, S. (eds) (2005) *Science, Technology and Society*, Blackwell Publishers, Malden, MA

Boin, A., Hart, P., Stern E. and Sundelius, B. (2007) *The Politics of Crisis Management: Public Leadership Under Pressure*, Cambridge University Press, Cambridge

Burby, R. J. (ed) (1998) *Cooperating with Nature: Confronting Natural Hazards with Land-Use Planning for Sustainable Communities*, Joseph Henry Press, Washington DC

Eglash, R., Croissant, J., Di Chiro, G. and Fouché, R. (eds) (2004) *Appropriating Technology: Vernacular Science and Social Power*, University of Minnesota Press, Minneapolis, MN

Fischer, F. (2000) *Citizens, Experts, and the Environment*, Duke University Press, Durham, NC

Frickel, S. and Moore, K. (2006) 'Prospects and challenges for a new political sociology of science', in Frickel, S. and Moore, K. (eds) *The New Political Sociology of Science*, University of Wisconsin Press, Madison, WI

Hess, D. (1997) *Science Studies: An Advanced Introduction*, NYU Press, New York

Hoffman, S. M. and Oliver-Smith, A. (2002) *Catastrophe and Culture: The Anthropology of Disaster*, School of American Research Advanced Seminar Series, Santa Fe, NM

Irwin, A. (1995) *Citizen Science: A Study of People, Expertise and Sustainable Development*, Routledge, New York

Irwin, A. and Wynne, B. (eds) (1996) *Misunderstanding Science? The Public Reconstruction of Science and Technology*, Cambridge University Press, New York

Kleinman, D. (2003) *Impure Cultures: University Biology and the World of Commerce*, University of Wisconsin Press, Madison, WI

Kleinman, D. (2005) *Science and Technology in Society*, Blackwell Publishing, Malden, MA

Martin, B. (ed) (1996) *Confronting the Experts*, SUNY Press, Albany, New York

Martin, B. (ed) (1999) *Technology and Public Participation*. University of Wollongong Press, Wollongong, Australia

Perry, R. W. and Quarantelli, E. L. (eds) (2005) *What is a Disaster? New Answers to Old Questions*, International Research Committee on Disasters, Xlibris Corporation, Philadelphia, PA

Phillips, B. D. (1993) 'Cultural diversity in disasters: Sheltering, housing, and long-term recovery', *International Journal of Mass Emergencies and Disasters*, vol 11, no 1, pp99–110

Rubin, C. B. (ed) (2007) *Emergency Management: The American Experience 1900–2005*, Public Entity Risk Institute, Fairfax, VA

Sclove, R. (1995) *Democracy and Technology*, Guilford Press, New York

Tierney, K. J., Lindell, M. K. and Perry, R. W. (eds) (2001) *Facing the Unexpected: Disaster Preparedness and Response in the United States*, Joseph Henry Press, Washington DC

Part I
Environmental, Cultural and Political Concerns

1
Katrina's Contamination: Regulatory Knowledge Gaps in the Making and Unmaking of Environmental Contention

Scott Frickel and M. Bess Vincent

Introduction

When Hurricane Katrina made landfall on 29 August 2005, its eye-wall grazed the south-eastern edge of the New Orleans metropolitan area, causing extensive wind and storm surge damage to built structures large and small and, most infamously, the concrete and earthen levee system surrounding the city. Overtopping and breeches in levee walls inundated 55 square miles of Orleans Parish with an estimated 131 billion gallons of flood water (Smith and Rowland, 2007; Campanella, 2010). The chemical- and pathogen-laced deluge soaked some parts of the city for nearly six weeks, creating an unprecedented risk scenario for state and federal agencies charged with assessing the storm's ecological and human impact.

In response, scientists from the US Environmental Protection Agency (EPA) and the State of Louisiana Department of Environmental Quality (DEQ) pursued extensive environmental testing of New Orleans flood sediment and soil. The hazard assessment programme spanned an entire year, involved collaborations with a dozen state and federal agencies and departments, and resulted in more than 400,000 tests for the presence of 195 substances, from about 1800 sediment and soil samples collected around the metropolitan area. EPA's final report, released in August 2006, claimed, as earlier reports also had, that 'the sediments left behind by the flooding from the hurricanes are not expected to cause adverse health impacts to individuals returning to New Orleans' (US EPA, 2006b).

Unlike the hapless Federal Emergency Management Agency (FEMA) response described by Rachel Dowty and colleagues (this volume), city and state officials, local health experts, business leaders, media outlets and residents alike found much to celebrate in EPA's apparent organizational competence in producing its environmental assessment. While a few area scientists, environmental justice and community activists (e.g. Nance, 2009), and at least one national environmental group have challenged some of EPA's claims (Solomon and Rotkin-Ellman, 2006), contention surrounding the official hazard assessment in New Orleans has, for the most part, been sporadic and short-lived. The relative lack of contention is perhaps not surprising, given the psychological, political and economic stakes faced by virtually every city resident struggling to rebuild their homes, their institutions and their lives in the wake of catastrophic loss. What we find interesting is the nature of those few challenges from outside experts and activists – nearly all of which focused on previously identified environmental hazards or on EPA's interpretation of some of their data. In stark contrast to the contentious nature of neighbourhood recovery planning efforts described by Barrios (this volume), EPA's challengers neither questioned how the agency produced its data, nor questioned what remained unknown.

Our research on Katrina's contamination asks how the absence of knowledge shapes contentious politics, and how contentious politics shapes the non-production of knowledge. Much has and continues to be written about the contested nature of scientific knowledge (Hackett et al, 2007) and a rapidly growing body of scholarship specifically targets the role of social movements in situating knowledge at the core of political and social struggle (e.g. Allen, 2003; Reardon, 2004; Moore, 2007). However, the contentious politics of 'undone science', in reference to marginalized, unfunded or generally ignored areas of research, has received comparatively little attention (Woodhouse et al, 2002; Hess, 2007; Frickel et al, 2010). The present study contributes to that slim body of work by developing the concept of 'knowledge gaps' and using that framework to examine undone EPA science in New Orleans.

One caveat: In describing the production of knowledge gaps in hazard assessment work, we do not mean to imply that people returning to live, work and play in New Orleans face exposure to dangerous chemical hazards (Picou, 2009). Our far more modest assertion is simply that, despite official assurances to the contrary, no one really knows what hazards remain.

What are knowledge gaps?

We conceptualize knowledge gaps as organizationally circumscribed domains of unrealized knowledge. Like knowledge itself, knowledge gaps are social products, the result of organizational practices, historical contingencies, and institutional arrangements that structure what we know and what we do not know. Our project is to understand how and under what conditions knowledge gaps are constituted and why some endure while others vanish, as it were, into substantive streams of knowledge practice.

This conceptualization owes much to 'undone science' (Hess, 2007). Discussions of undone science situate knowledge production within the institutional matrix of state, industry and social movements. Most of this research to date emphasizes the role of social movements in drawing attention to underdeveloped streams of research and the challenges allied scientists often face in pursuing that research. In its emphasis on marginalized knowledge fields, undone science serves as an orienting concept for the present project. We see knowledge gaps as organizational outcomes of undone science.

Knowledge gaps demarcate absences. They describe negative conditions that signal a lack of knowers and consequently a lack of ideas, practices and networks. Since these are the resources that drive knowledge systems, knowledge gaps mark certain domains as inaccessible. In the language of actor-network theory (Latour, 1987), knowledge gaps are domains unpopulated by actants, where enrolment becomes a practical impossibility because there is nothing or no one to enrol. In the language of social network theory, knowledge gaps are akin to structural holes (Burt, 1992); they delineate paths of greatest resistance and in so doing channel resources toward paths of lesser resistance. The presence of these absences (to use an odd phrasing) passively reinforces existing institutional arrangements by leaving the status quo unquestioned and by leaving questions about alternatives unasked. In this sense, knowledge gaps function as latent structures. They facilitate possibilities for knowledge in certain organizational domains and networks by foreclosing possibilities in others.

Characterized thus, knowledge gaps are social products and have social effects, both of which are, in principle, identifiable and measurable. But knowledge gaps are also social processes; they are emergent phenomena, shaped by human interactions, organizational practices, and institutional pressures past and present. And they are highly variable. However, we propose that knowledge gaps are not randomly distributed. Thus, in addition to identifying and describing their social features and effects, the sociological analysis of knowledge gaps should also seek to explain their distribution across knowledge systems in relation to the processes that condition their emergence.

We also see knowledge gaps as distinct from questions of risk. Risk is an institutional response to scientific uncertainty in regulatory settings in which the statistical application of probability theory is used to stabilize and manage knowledge claims (Beck, 1999; Van Loon, 2002). By stabilizing knowledge within a range of possible outcomes deemed more or less likely, risk analysis purports to provide a rational basis for objective decision making under conditions of controlled uncertainty (Dietz and Rycroft, 1987; Jaeger et al, 2001).

This distinction gains importance in our case study, where some environmental organizations working in New Orleans have contested the standards EPA has used to make risk claims. In contrast, the agency's prior task of identifying hazardous substances – which produces the knowledge on which risk assessments are based – has not been the subject of public debate. It is our subject, however, because claims of hazard identification rest on prior decisions about what will be known and what will remain unknown.

Studying what is not there

An empirical research programme to study knowledge gaps is in many ways consonant with Woodhouse's call (this volume) for a symmetrical perspective on socio-technical outcomes. In our case, however, extending Bloor's (1976) symmetry principle to investigate what is known/unknown requires figuring out how to study something that is not there. Knowledge gaps leave faint traces that can be difficult to track, especially in the absence of social movements organized to make those lacunae visible to the outside world (e.g. Hess, 2007). Even then, systematic data on what does not exist (or exists only minimally) is rare, and scholars are often forced by lack of contiguous chains of evidence to develop arguments based in part, at least, on counterfactual claims. In light of these challenges, we focus our attention on EPA's organizational domain. Doing so takes advantage of the bureaucratic rules and rituals that govern scientific work in regulatory contexts and demonstrates the viability of our approach for making knowledge gaps visible.

In the aftermath of an urban-scale disaster such as occurred in New Orleans, the ideal strategy for minimizing gaps in knowledge about environmental hazards is one that is uniformly comprehensive. Such an approach would sample as much as possible in as many places and times as possible and test each of those samples for as many chemicals as possible – a strategy that is clearly unrealistic in practical terms. But it suggests that a relatively straightforward way to identify knowledge gaps is to pay close attention to how uniformity and comprehensiveness was produced within and across efforts to collect and analyse data. The EPA test result data allow us to do this, not by examining test results per se, but by examining when, where and how often samples were collected, and how many of what kind of analyses were conducted on the sampled material. Then we look for variation and patterns. In theory, knowing how sampling and chemical analysis was organized will tell us something about the spatial, temporal and institutional limits of EPA's knowledge about contamination. Where we find the limits of knowledge we should also find the edges of knowledge gaps.

For this we rely on data generated by EPA's Region 6 Katrina and Rita Hurricane Response Project.[1] Many of these are publicly available and include published summary reports, press releases, public correspondence and, most importantly, the EPA environmental test results data (available at www.epa.gov/katrina/index.html). The database contains parameters and results for every chemical and biological analysis conducted on every sample collected by EPA in the hurricane-impacted region from September 2005 to June 2006. Each record corresponds to the results of a single test for the presence of a single compound. Key variables for our study include date, time, longitude/latitude coordinates, unique activity and sample IDs, industrial chemical code, analyte name and environmental media (soil or sediment). By aggregating and comparing these variables, we are able to show what kinds of potential contaminants were investigated when, how and where.

Limited additional information has been provided by four informants (two from EPA and two who have closely monitored the project), as well as a small number of internal documents obtained from EPA through a Freedom of Information Act request.

New Orleans hazard assessment

As described in various summary reports (US EPA, 2005a, 2005b, 2006a, 2006b), the EPA-led hazard assessment project in New Orleans involved four distinct phases of data collection and laboratory analysis. Unless otherwise noted, information presented in this section is derived from these reports and from our analysis of the EPA test results data.

Phase I

Sampling for Phase I began on 10 September 2005, nearly two weeks after Katrina struck, and ended on 14 October. Regulators sought to identify hazardous substances in residential area sediment and assess whether those substances posed any short-term exposure risks to emergency workers and residents. Based on the general assumption that flood waters would distribute suspended environmental hazards uniformly, EPA field teams collected samples from evenly spaced geographical points across the city to characterize the

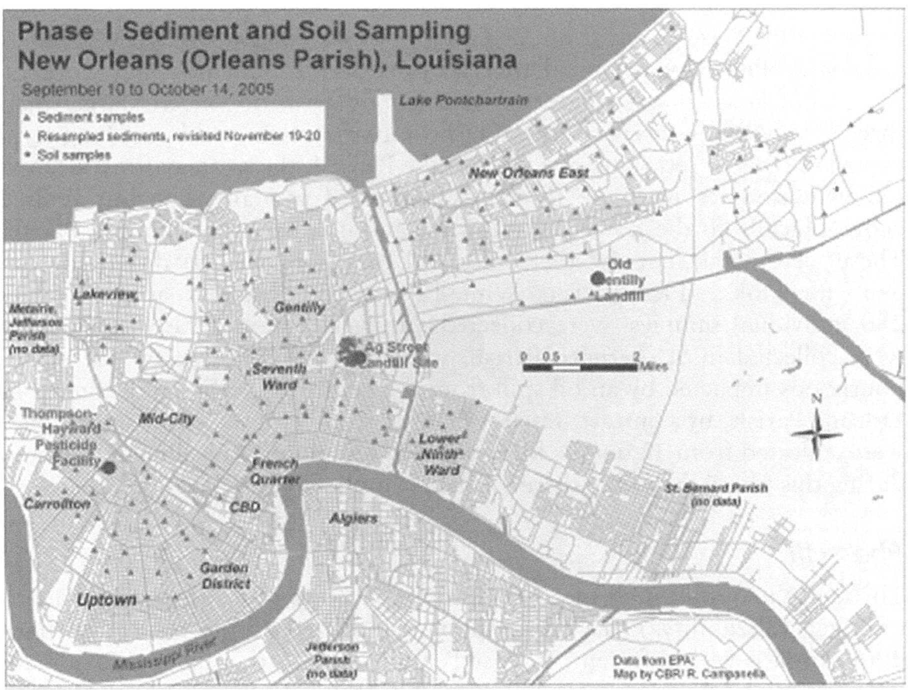

Source: Data from EPA; map by R. Campanella

Figure 1.1 *Phase I samples*

chemical content of flood sediment (telephone interview, 13 September 2006). About 450 samples were collected overall. In Orleans Parish, collection teams took 239 sediment samples from 215 locations.[2] Once collected, 2 commercial laboratories analysed these samples for the presence of 195 different chemical analytes and *E. coli* bacteria.

This method identified 145 locations across the metropolitan region where contaminants were found at sufficient levels to require further investigation. These sites were targeted for resampling, which took place a month later (19–20 November 2006) in the midst of what would become Phase II. By then, the effects of wind, rain, traffic, street repair and debris removal operations had substantively altered the topography of sediment. Collection teams found sediment at sufficient depths for resampling (i.e. 0.5cm or greater) at only 14 of these locations. Eleven of the 14 sample sites were in Orleans Parish.

In addition to samples of surface sediment, 23 below-surface soil samples were collected during Phase I, all of them at or near the former Agriculture Street Landfill Superfund site. This is a hazardous waste site that, since the early 1990s, has been a focus of environmental justice struggles and sporadic remediation efforts (Roberts and Toffolon-Weiss, 2001; Colten, 2005). The samples collected at this location were tested for arsenic and lead only.

Phase II

Sampling in Phase II ran from 29 October to 27 November 2005. Whereas the goal of Phase I was immediate risks to residents and emergency workers, the goal of Phase II was to identify generalized 'areas of concern' larger than three acres where concentrations of hazardous materials could pose long-term exposure risks to returning residents.[3] Like Phase I, the strategy for collecting sediment samples in Phase II was systematized spatially, but was limited geographically to the Lower Ninth Ward and in adjacent St Bernard Parish. These were areas that, by this time, had been reopened to displaced residents for 'look and leave' visits. While EPA reports note that 'approximately 280 individual samples' were collected during this phase, the vast majority were collected in St. Bernard Parish, where concern focused on the neighbourhoods impacted by an oil spill from the nearby Murphy Oil Refinery. In Orleans Parish, by contrast, test result data show that 41 sediment samples were collected from 32 unique locations. No resampling of sediment occurred during this phase, nor were there any soil samples collected.

Phase III

Through the autumn/fall of 2005, sampling efforts were designed to maximize generalizability. Phase III sampling, conducted during 16–22 February 2006, took a very different, 'focused' approach. In this phase EPA's main goal was to characterize spatial plumes of contamination at 32 New Orleans locations where earlier testing had indicated high concentrations of arsenic, lead or benzo(a)pyrene (a known carcinogen and a constituent component of petroleum fuels).

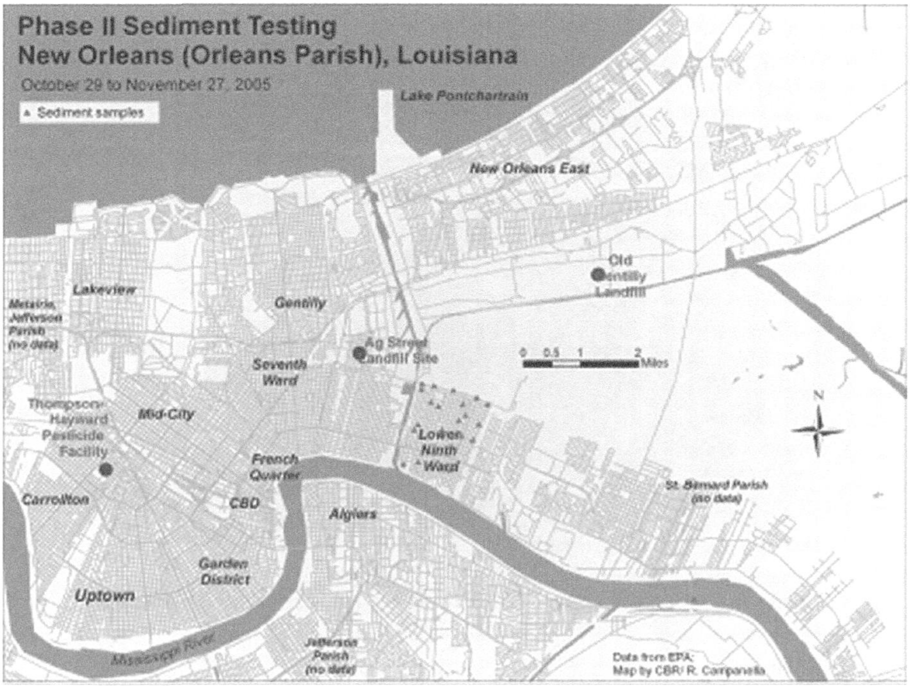

Source: Data from EPA; map by R. Campanella

Figure 1.2 *Phase II samples*

At each of these 32 sites, EPA collected 4 samples of 'sediment impacted soil' from each of several points arrayed within a 500ft radius around the original sample point. In the lab, researchers combined the clustered samples using a 'composite' protocol to homogenize the media and then tested the combined samples for one or more of the three analytes in question. Using this method, EPA field crews collected 256 soil samples from 225 locations, all in close proximity to one of the 32 'areas of concern' identified during Phases I and II.

In response to community and activist concerns, assessment teams also collected another 46 samples from the Agriculture Street Landfill site, 13 samples from a Seventh-Ward residential neighbourhood surrounded by railroad tracks, warehouses and highway interchanges, and 10 samples near an abandoned pesticide blending plant, once operated by the Thompson-Hayward Chemical Company. Like the Agriculture Street Landfill site, the old Thompson-Hayward property had attracted neighbourhood opposition and EPA attention after the facility closed in 1988 (Schleifstein, 1995). Lab testing focused on 127 analytes at the Agriculture Street and Seventh-Ward sites, and 23 analytes at Thompson-Hayward.

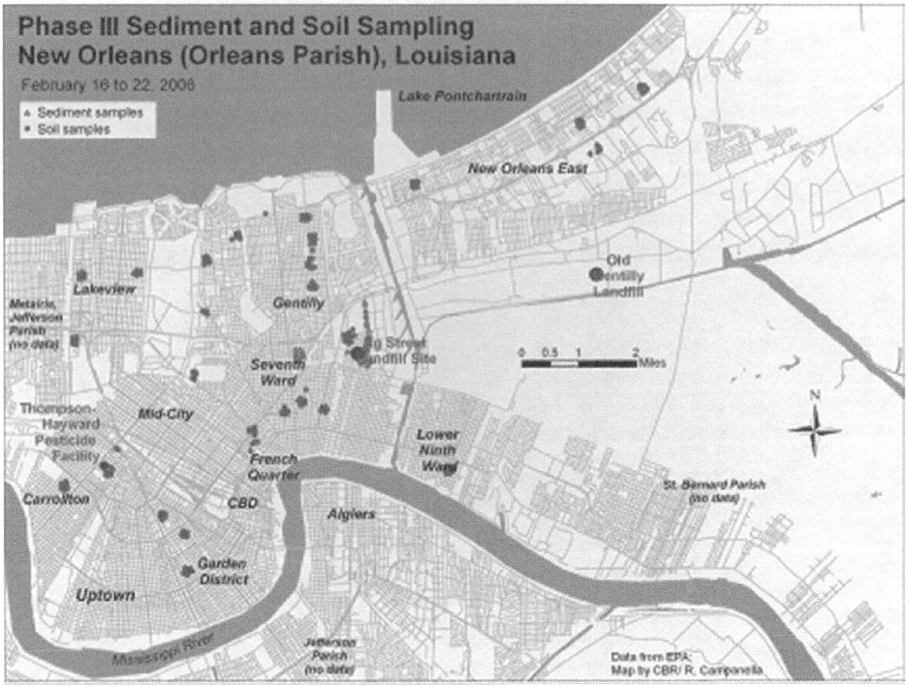

Source: Data from EPA; map by R. Campanella

Figure 1.3 *Phase III samples*

Phase IV

Phase IV samples were collected in New Orleans in April and early May, seven months after Katrina struck. By then, EPA goals had again shifted. This time, researchers' objective was to locate and analyse remaining flood sediments to update analyses from Phases I and II. Like these earlier efforts, Phase IV sampling was systematic and like Phase II, it targeted a specific part of the city. This time, a denser grid based on 200ft interval points was mapped across the eastern third of New Orleans and the adjacent city of Chalmette, in neighbouring St. Bernard Parish. Field crews collected 464 samples from 390 eastern New Orleans locations. As noted in a summary report, samples were *not* collected from more than 1000 pre-selected grid points for lack of sediment or because those sites were located in non-residential areas. Ninety-five per cent of the samples were tested for 124–128 analytes; the remaining 5 per cent were tested for 193. Finally, in response to more demands from concerned citizen groups and environmental organizations, field crews collected 11 soil samples from the Old Gentilly Landfill, a closed landfill that city officials reopened after Katrina to take storm debris. These 11 samples were tested for 70 analytes.

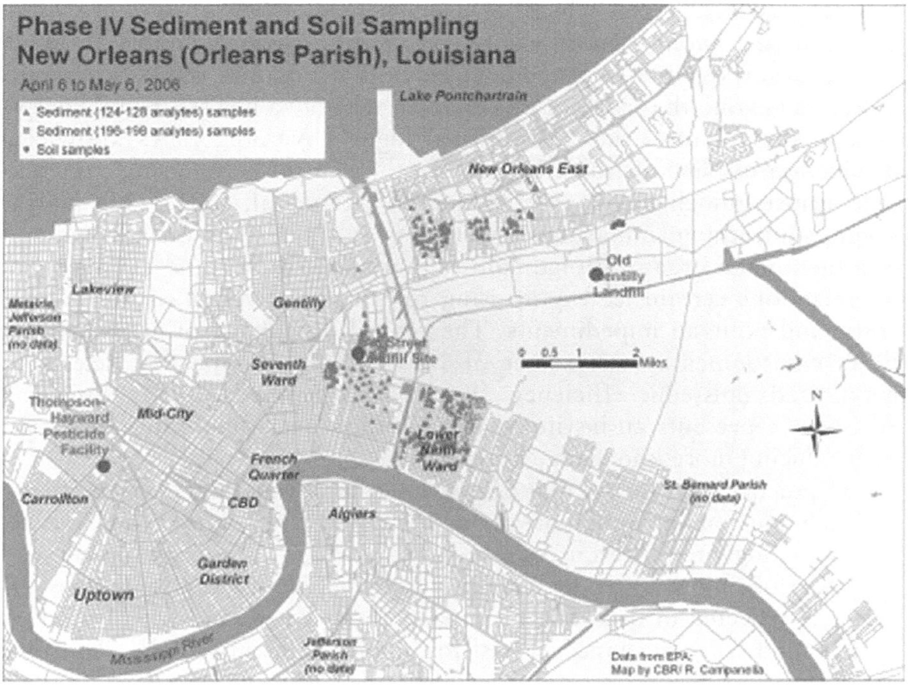

Source: Data from EPA; map by R. Campanella

Figure 1.4 *Phase IV samples*

The institutional logic of knowledge gap production

For many residents and city leaders, EPA's hazard assessment project effectively put to bed questions about the lingering health impacts of the 'toxic soup' presumably cooked up by Hurricane Katrina. It did so by producing a wall of data comprised of more than 400,000 tests showing that, but for two or three specific substances in a handful of specific locations, contaminant levels in New Orleans flood sediment did not exceed federal or state health screening levels for chemical toxicants. The assessment results have been vigorously defended by collaborating organizations (Sinks, 2006) and the findings reconfirmed by teams of academic and industry scientists (Marris, 2006). Perhaps most importantly, EPA's innovative 'compressed risk management approach' is being promoted as a model for future regulatory responses to large-scale catastrophes (Walsh et al, 2006). Thus, it is all the more important to understand the social, organizational and epistemic limits that this approach to knowledge production institutionalizes (Frickel and Vincent, 2007).

As we interpret it, the sampling and testing regime developed by EPA is a product of what Karin Knorr Cetina calls 'epistemic cultures: those amalgams of arrangements and mechanisms – bonded through affinity, necessity, and historical coincidence – which, in a given field, *make up how we know and what we know*' (Knorr Cetina, 1999, p1, original emphasis). While detailed analysis of the epistemic cultures that organize scientific practice at EPA is beyond the scope of this chapter, we note a key structural component of that culture: the 'institutional logic' governing EPA's knowledge production in New Orleans, by which we mean the 'organizing principles that furnish guidelines for practical action' (Rao et al, 2005). Our analysis suggests intriguing similarities with Dowty et al's description (this volume) of a hierarchist organizational culture that allowed EPA and collaborating organizations certain leeway in using rules to creatively navigate bureaucratic and political impediments. The central organizing principle guiding EPA's environmental assessment of New Orleans is one that selectively maximized 'epistemic efficiency' (Frickel and Edwards, 2011). Epistemic efficiency represents an institutional logic that, over time, rationalizes risk by producing more knowledge about less.

In practice, this meant that EPA focused increasing money, time, energy and technique on knowing more about what was already known and invested correspondingly fewer resources in examining what was not already known – and therefore more likely to remain unknown. The logic of epistemic efficiency is reflected in decisions EPA made about where and when to collect samples and what substances to test for among different samples. This logic also seems to have guided decisions about whether and where to resample environmental media or retest sampled material. While there are exceptions, in general the EPA hazard assessment project consistently produced more knowledge about fewer substances present in fewer New Orleans neighbourhoods. One outcome of the consistent application of this institutional logic is the generation of knowledge gaps with distinctly socio-spatial dimensions.

The map presented in Figure 1.5 illustrates the spatial organization of knowledge and knowledge gaps in relation to the Katrina floodprint. Before discussing this topic, two related issues warrant brief mention. First, the map clearly shows that, in line with its institutional mandate to investigate chemical hazards distributed by the flood waters, EPA did not collect any samples in parts of the city that did not flood. This may prove significant in light of the fact that the city's 'high ground', particularly along the Mississippi River, is where industry has concentrated historically and is where soil contamination is likely to be present. After Katrina, the old warehouse and manufacturing district near the city's downtown central business district (CBD) has become prime real estate targeted for intensive residential redevelopment (Campanella, 2007).

Second, city topography is such that flood waters were likely not as uniform as assumed by agency experts. The below sea-level 'bowl' of New Orleans is trisected by three ridges, the geological remains of natural levees created over thousands of years as the Mississippi River slowly shifted course across its delta (see also Bowden, this volume). The Metarie and Gentilly Ridges run roughly east–west while the Esplanade Ridge runs from the river northward, all more or less converging near the residential heart of New Orleans in Mid City and helping to form the cities' distinct hydrological sub-basins. Since

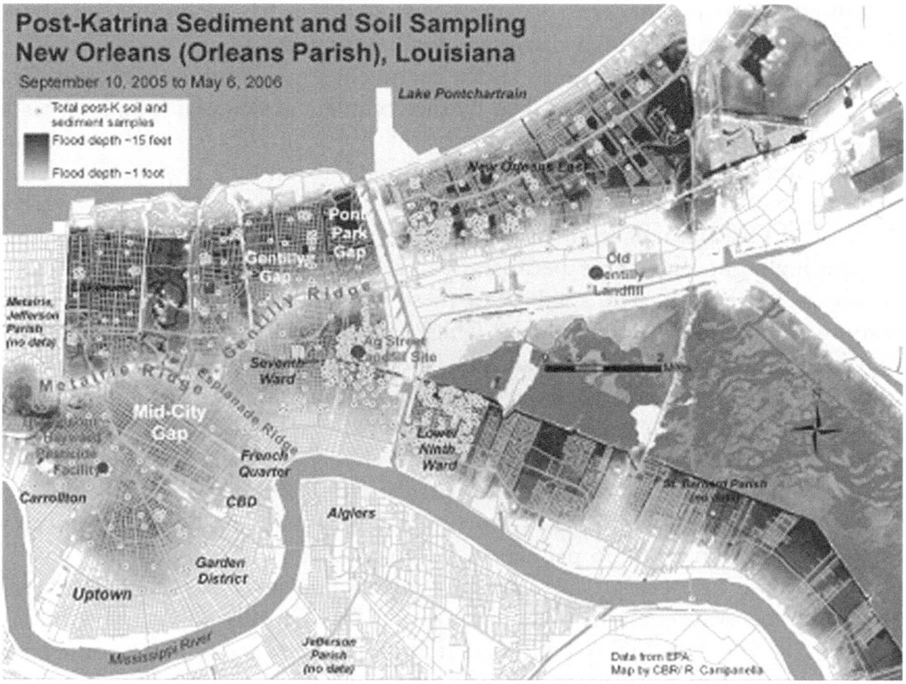

Source: Data from EPA; map by R. Campanella

Figure 1.5 *Total samples*

the flood data shown in Figure 1.5 indicate that, for the most part, water did not cross these ridges, we can be fairly certain that in chemical terms the water that drowned homes in Gentilly or Pontchartrain Park neighbourhoods was different from the water that drowned homes in Mid City neighbourhoods. As carried out in Phases I and II, EPA's strategy to provide a generalized assessment of flood-deposited contaminants assumed the flood represented a uniform hazard and ignored the hydrological effects of urban topography.

The key features of our analysis, however, are knowledge gaps. The dots populating Figure 1.5 represent the aggregate results of sample collection in New Orleans in 2005–2006. Readers should keep in mind that the relative size of these dots – scaled to roughly one city block – is a cartographic artifact; in fact, they each represent less than three cubic inches of organic material. Where sampling points were distributed fairly evenly across the city in Phase I (refer to Figure 1.1), by the end of the project this distribution had become highly uneven. This is because the logic of epistemic efficiency concentrated the production of site-specific knowledge primarily in the neighbourhoods surrounding the old Agriculture Street Landfill, in the Lower Ninth Ward, and in neighbourhoods east of the Industrial Canal. Smaller clusters in other parts of the city indicate where 'focused' sampling to further investigate previously identified 'areas of concern' occurred during Phase III, including the abandoned Thompson-Hayward chemical blending plant.

We identify 'knowledge gaps' as the relatively large contiguous areas within the floodprint that are distinguished by the complete absence of sampling. Such gaps represent epistemologically significant holes in the hazard assessment project, and are one way in which our approach to undone regulatory science makes knowledge gaps visible. Figure 1.5 shows three such gaps that overlay three New Orleans neighbourhoods. These are not the only knowledge gaps, but we highlight them here because their relative size and shape make them visibly more obvious. The smallest of these we call the Gentilly gap, which is composed of roughly 30 city blocks. The Pontchartrain Park gap is next in size, representing about 155 city blocks. The Mid City gap is the largest of the three, encompassing about 250 blocks. All three gaps intersect densely populated residential areas, and their pre-Katrina population characteristics illustrate some of the city's racial and socio-economic diversity. These gaps span census tracts with pre-Katrina populations of African-American residents above and below the city-wide average and with the proportion of residents living below the poverty level ranging from 10 to 56 per cent (US Census, 2000).

The three knowledge gaps are also distinctive in terms of their potential for concentrating environmental hazards. The Gentilly gap is immediately adjacent to the London Avenue Canal breech, an event which produced a plume of sediment up to seven feet deep and extending for several blocks from the mouth of the breech (Nelson and Leclair, 2006). The Pontchartrain Park gap, adjacent to the west edge of the Industrial Canal, was also not sampled by EPA teams. While EPA traced several instances of arsenic contamination to herbicides used to maintain nearby golf courses in other parts of the city (McDaniel, 2006), no information was collected from sediment at residences adjacent to the public golf course at the centre of the Pontchartrain Park neighbourhood. The Mid City gap, although mostly residential, was also home to a heavily industrialized area extending more than a mile from the central business district (CBD). As we know from related research (Frickel and Elliott, 2008), at least a dozen hazardous industrial facilities have been located in the heart of this knowledge gap over the last 50 years.

These knowledge gaps are spatially realized forms of undone regulatory science – the institutionalized outcomes of EPA knowledge production systems organized by a logic of epistemic efficiency. Yet the present analysis only begins to describe their spatial character and much work remains. For example, future research will formalize a process for quantitatively identifying knowledge gaps based on the variable distances between sample points, using the average point-to-point distances from Phase I samples for a baseline. Another avenue open to further investigation involves the public vs. private nature of the properties where sampling occurred. The vast majority of samples were collected by EPA on public property, primarily from street gutters, medians and sidewalks.[4] Thus at a finer scale of measurement, the project produced little or no place-specific knowledge about contaminants in residents' back yards or inside their homes. Such 'backyard' knowledge gaps represent a politically significant shift in accountability from the state to private citizens, as they place the burden of knowledge production and risk assessment on individual homeowners. Future research will need to more comprehensively address the spatial dimensionality of undone regulatory science, as well as the temporal, organizational and other dimensions of knowledge gap production. This work is ongoing (see Frickel and Vincent, 2007; Frickel et al, 2009; Frickel and Edwards, 2011). For now, we turn to a consideration of the interplay between knowledge gaps and environmental contention.

How environmental contention conditions knowledge gaps

In describing the formation of knowledge gaps as one element of undone regulatory science, our analysis has emphasized the institutional logic of knowledge production. In that project context, what passed as controversy was largely contained within the formal organizational and communication networks that defined the multi-agency effort. Yet there were moments when controversy became quite public, involving concerned city and parish officials, local experts, community groups and environmental organizations. Such public dissent has indirectly shaped the non-production of knowledge in New Orleans in at least three ways.

First, public controversy shaped what became known. During autumn/fall 2005, community and environmental groups demanded that EPA intensify testing efforts in New Orleans neighbourhoods known to be contaminated before Katrina struck. Specifically, these groups asked EPA to conduct additional sampling and testing at the Agriculture Street Landfill Superfund site, the Thompson-Hayward facility, and at the Old Gentilly landfill. Activists also demanded that the Lower Ninth Ward receive additional regulatory attention after sediment from the pollution-laced Industrial Canal silted through the levee breech there. EPA responded to these demands with more sampling and more testing in areas that, while known to be contaminated, would not likely have received such increased attention without community and NGO pressure (US EPA, 2006b).[5]

Contention also emerged around EPA's initial report on the hazard assessment results. In December 2005, EPA released a report summarizing its conclusions regarding areas of metropolitan New Orleans impacted by the flooding (US EPA, 2006a). The report was accompanied by a DEQ press release stating that returning residents faced 'no unacceptable health risks' (2005). In response, the Natural Resources Defense Council issued a report that was highly critical of EPA strategies, suggesting that EPA selectively used different state and federal screening standards to assess health risks (Solomon and Rotkin-Ellman, 2006). The ensuing debate occurred through a series of public letters, and focused on legal questions concerning appropriate screening limits and regulatory action derived from test results. (The exchange is available at www.deq.louisiana.gov/portal/tabid/2245/Default.aspx.)

A third way contention shaped the hazard assessments involved the way information was conveyed to the general public. In the wake of the flooding, community groups, environmental organizations and government officials from the impacted parishes insisted that the test results be made available to the public in language that non-experts could understand. EPA provided that information in two forms. One was in the test results database, which presents knowledge as raw data organized by individual test results. This strategy makes data accessible but without meaningful interpretations. Even technically competent outsiders had difficulty navigating through the data and aggregating it in meaningful ways (Informant interview, 23 July 2007). EPA also made information available through summary assessment reports. As requested by citizens, the reports were brief and written in non-technical language that lay persons could understand. However, these reports contained very little useful place-specific information about particular neighbourhoods or blocks. While raw data were uninformative because there was no spatial aggregation of test results, the summary reports were uninformative because they condensed too much information to be useful to people worried about potential hazards in their neighbourhood or on their street. Instead, the summary assessment reports reassured concerned citizens that the regulatory agencies charged with protecting the public from storm-related environmental hazards were doing just that, and that the evidence the hazard assessment had produced showed little cause for further concern. The reports neither addressed the limits of that evidence nor acknowledged questions that remained unanswered.

These three examples exemplify the dynamic interplay between knowledge gaps and environmental contention. However, sporadic, public dissent was instrumental in shaping what became known, what that knowledge meant, and how that meaning was conveyed to the public. These same moments of environmental contention also contributed to the production of knowledge gaps by generating public demand for environmental knowledge in ways that reinforced EPA's institutional logic of epistemic efficiency. Activists' demands for more knowledge and more transparent communication of that knowledge channelled agency resources toward deepening existing knowledge – for example, about the nature of contamination at the Agriculture Street Landfill site – rather than working to narrow knowledge gaps by extending knowledge production to other previously neglected neighbourhoods.

This shaping process is bidirectional. The resulting knowledge gaps also exert considerable, if latent and largely unrecognized, influence on public demands for additional environmental knowledge. In a city that remains desperate to regain its economic and demographic footing, much of what was left unknown in the autumn of 2005 remains unknown today. Quiescence reigns in the public sphere on the question of long-term health risks from the city's contaminated soils, not because there is nothing left to learn, or because people necessarily trust the official assessments, but because knowledge gaps have altered the terrain of permissible questions. In the context of a city that continues to undergo massive ecological and urban transformation, institutionalized unknowns quickly become 'unknowable' (Hilgartner, 2001). When that happens, as it may already have in New Orleans, challenges to undone EPA science pass out of the realm of the politically possible.

Concluding observations

This chapter investigates the social production of knowledge gaps, using the case of EPA's hazard assessment in post-Katrina New Orleans to illustrate that process. Our analysis points to five observations about knowledge gaps in regulatory contexts. The first is that knowledge gaps are not metaphorical abstractions but identifiable and measurable outcomes of undone science. The complex of bureaucratic rules, regulations, standards and protocols used to produce regulatory knowledge can also be used strategically to render knowledge gaps visible and thus amenable to social analysis and assessment.

A second observation is that knowledge gaps are multidimensional. Although the analysis presented in this chapter has stressed their spatial configuration, knowledge gaps may be better understood as spatial outcomes produced by distinctly temporal, organizational and ecological processes. What became known and what remained unknown depended not only on where samples were collected, but when and under what environmental and social conditions.

A related observation is that knowledge gaps are also products of distinct epistemic cultures. An important element of epistemic culture is the institutional logic through which scientific organizations identify, gather, organize and interpret data. In the case presented here, EPA developed a knowledge production system designed to maximize epistemic efficiency. This involved accumulating and concentrating the agency's hazard assessment and public communication efforts on evidence that required regulatory action, but ignoring evidence when contaminant levels were below regulatory limits, and ignoring neighbourhoods distinguished by the absence of state-generated evidence.

Our fourth observation is that activist and other public groups play vital if often unrecognized roles in shaping decisions about what will be known and what will remain unknown. As existing research shows, social movements have the potential to actively narrow existing knowledge gaps by drawing attention to undone science. This study additionally suggests that social movements and their expert allies may also widen or deepen existing knowledge gaps unintentionally. This occurs because the questions activists ask are often derived from the answers science has previously provided. In this way, knowledge gaps latently and indirectly influence the form and content of strategic collective action that social movements undertake.

A fifth observation is that knowledge gaps have life cycles with historically traceable beginnings and endings. One way knowledge gaps disappear is through coordinated efforts to answer the questions they contain. But another way, which seems to emerge from the present analysis, is to do nothing. Particularly with respect to spatially specific time-sensitive knowledge, regulatory knowledge gaps may disappear through bureaucratic processes of knowledge attrition in the face of rapidly changing ecological and political contexts.

Together, these observations provide a starting point for theorizing the social nature of knowledge gaps and for developing an empirical programme to investigate and explain their emergence and their social effects. The broader political significance of such a programme lies in the capacity of that research to illuminate tensions and contradictions in the institutional logics that govern and concentrate knowledge production inside the state, where decisions about what knowledge is needed are made by the same organizations that decide whether and how that knowledge is used.

Acknowledgements

This chapter was originally presented at the Seminar on Contentious Knowledge: Science, Social Science, and Social Movements at the Institute for the Social Sciences, Cornell University. We thank seminar participants Maria Cook, Durba Ghosh, Ron Herring, Stephen Hilgartner, Ken Roberts and Sid Tarrow for critical consideration of our ideas. Gene Rosa and Michelle Edwards also provided insightful comments. We also thank Richard Campanella for producing the maps, and the editors for excellent copy-editing. We alone are responsible for whatever errors remain. Resources for research and writing time were graciously provided through the Tulane/Xavier Center for Bioenvironmental Research and by grants from the Social Sciences Research Council and the Southern Sociological Society.

Notes

1 The EPA-led hazard assessment was not the only environmental testing to occur in metropolitan New Orleans or in the hurricane-impacted region more generally. In the aftermath of the storm and flooding several local and national environmental groups, including the Louisiana Bucket Brigade, the Louisiana Environmental Action Network, the Natural Resources Defence Council and the Sierra Club (Delta Chapter) were involved in various smaller-scale environmental testing efforts.
2 Under EPA quality assurance protocol, two samples must be collected at 10–20 per cent of sample collection locations. Each duplicate sample is then sent to different laboratories for analysis. This duplicate sampling accounts for differences between samples collected and sample locations.
3 EPA defines 'areas of concern' as locations where lifetime cancer risk exceeds 1 in 10,000.
4 The main exceptions were homes in the Agriculture Street Landfill neighbourhood (Informant interview, 2 March 2006).
5 It is likely that the media's extended focus on the devastation in the Lower Ninth Ward also contributed to the regulatory response to that area.

References

Allen, B. (2003) *Uneasy Alchemy: Citizens and Experts in Louisiana's Chemical Corridor Disputes*, MIT Press, Cambridge, MA

Beck, U. (1999) *World Risk Society*, Polity Press, Cambridge, UK

Bloor, D. (1976) *Knowledge and Social Imagery*, University of Chicago Press, Chicago, IL

Burt, R. S. (1992) *Structural Holes: The Social Structure of Competition*, Harvard University Press, Cambridge, MA

Campanella, R. (2007) 'Above-sea-level New Orleans: The residential capacity of Orleans Parish's higher ground', http://kerrn.org/pdf/campanellaaslno.pdf

Campanella, R. (2010) Personal communication, 12 January
Colten, C. E. (2005) *Unnatural Metropolis: Wresting New Orleans from Nature*, Louisiana State University Press, Baton Rouge, LA
Dietz, T. M. and Rycroft, R. W. (1987) *The Risk Professionals*, Russell Sage, New York
Frickel, S. and Edwards, M. (2011) 'Untangling ignorance in environmental risk assessment', in N. Jas and S. Boudia (eds) *Powerless Science? The Making of the Toxic World in the Twentieth Century*, Berghahn Books, Oxford and New York, forthcoming
Frickel, S. and Elliott, J. R. (2008) 'Tracking industrial land use conversions: A new approach for studying relict waste and urban development', *Organization & Environment*, vol 21, no 2, pp128–147
Frickel, S. and Vincent, M. (2007) 'Katrina, contamination, and the unintended organization of ignorance', *Technology in Society*, vol 29, pp181–188
Frickel, S., Gibbon, S., Howard, J., Kempner, J., Ottinger, G. and Hess, D. (2010) 'Undone science: Charting social movement and civil society challenges to research agenda setting', *Science, Technology & Human Values*, vol 35, no 4, pp444–473
Hackett, E. J., Amsterdamska, O., Lynch, M. and Wajcman, J. (eds) (2007) *The New Handbook of Science and Technology Studies*, MIT Press, Cambridge, MA
Hess, D. J. (2007) *Alternative Pathways in Science and Industry: Activism, Innovation, and the Environment in an Era of Globalization*, MIT Press, Cambridge, MA
Hilgartner, S. (2001) 'Election 2000 and the production of the unknowable', *Social Studies of Science*, vol 31, pp439–441
Jaeger, C. C., Renn, O., Rosa, E. A. and Webler, T. (2001) *Risk, Uncertainty, and Rational Action*, Earthscan, London
Knorr Cetina, K. (1999) *Epistemic Cultures: How the Sciences Make Knowledge*, Harvard University Press, Cambridge, MA
Krupa, M. and Donze, F. (2007) 'Leaders stand united behind recovery plan: Blueprint addresses fears of abandonment', *The Times Picayune*, 30 March, National section, Friday online edition, http://www.nola.com/news/t-p/frontpage/index.ssf?/base/news-8/1175234089230740.xml&coll=1
Latour, B. (1987) *Science in Action*, Open University Press, Milton Keynes
Louisiana Department of Environmental Quality (2005) 'State, federal agencies summarize environmental assessment of hurricane impacted areas of Southeast La', press release, Louisiana Department of Environmental Quality', www.deq.louisiana.gov/portal/portals/0/news/pdf/jointenvironmentalassessmentpr.pdf, accessed 10 December 2005
Marris, E. (2006) 'New Orleans cleared of "toxic soup" scenario: Surveys show no evidence of long-term health risks caused by Katrina', *Nature News*, www.nature.com/news/2006/060911/full/060911-14.html, accessed 15 September 2006
McDaniel, M. D. (2006) 'Letter to Patrice Simms', www.deq.louisiana.gov/portal/portals/0/news/pdf/DEQEnvAssessmentKatrinaNRDCResponse.pdf, accessed 2 February 2006
Moore, K. (2007) *Disruptive Science: Social Movements, American Scientists, and the Politics of the Military, 1945–1975*, Princeton University Press, Princeton, NJ
Nance, E. (2009) 'Making the case for community-based laboratories: New strategies in the environmental justice movement', in R. Bullard and B. Wright (eds) *Race, Place, and Environmental Justice After Katrina*, Westview Press, Boulder, CO, pp153–166
Nelson, S. A. and Leclair, S. F. (2006) 'Katrina's unique splay of deposits in a New Orleans neighbourhood', *GSA Today*, vol 16, pp4–9
Picou, S. J. (2009) 'Katrina as a natech disaster: Toxic contamination and long-term risks for residents of New Orleans', *Journal of Applied Social Sciences*, Fall, pp39–55

Rao, H., Monin, P. and Durand, R. (2005) 'Border crossing: Bricolage and the erosion of categorical boundaries in French gastronomy', *American Sociological Review*, vol 70, no 9, pp968–991

Reardon, J. (2004) *Racing to the Finish: Identity and Governance in an Age of Genomics*, Princeton University Press, Princeton, NJ

Roberts, J. T., and Toffolon-Weiss, M. M. (2001) *Chronicles from the Environmental Justice Frontline*, Cambridge University Press, New York

Schleifstein, M. (1995) 'Danger at the doorstep', *The Times Picayune*, 29 January, A1

Sinks, T. H. (2006) 'Letter and addendum to Patrice Simms', www.deq.louisiana.gov/portal/portals/0/news/pdf/ATSDRlettertoNRDC-080206.pdf, accessed 2 August 2006

Solomon, G. M. and Rotkin-Ellman, M. (2006) *Contaminants in New Orleans Sediment: An Analysis of EPA Data (February)*, Natural Resources Defense Council, Washington DC

Smith, J. and Rowland, J. (2007) 'Temporal analysis of floodwater volumes in New Orleans after Hurricane Katrina', in G. S. Farris, G. J. Smith, M. P. Crane, C. R. Demas, L. L. Robbins and D. L. Lavoie (eds) *Science and the Storm: The USGS Response to the Hurricanes of 2005*, pp57–61

US Environmental Protection Agency (2005a) 'Environmental assessment summary for areas of Jefferson, Orleans, St. Bernard, and Plaquemines Parishes flooded as a result of hurricane Katrina', www.epa.gov/katrina/testresults/katrina_env_assessment_summary.htm, accessed 6 December 2005

US Environmental Protection Agency (2005b) 'Summary of sediment testing from September 10–November 27, 2005: Hurricanes Katrina and Rita', www.epa.gov/katrina/testresults/sediments/index.html, accessed 2 December 2005

US Environmental Protection Agency (2006a) 'Summary assessment of the results of sampling of localized areas identified for focused investigations following hurricane Katrina', www.epa.gov/katrina/testresults/sediments/focused_sampling.html, accessed 31 March 2006

US Environmental Protection Agency (2006b) 'Summary results of sediment sampling conducted by the Environmental Protection Agency in response to hurricanes Katrina and Rita', www.epa.gov/katrina/testresults/sediments/summary.html, accessed 17 August 2006

Van Loon, J. (2002) *Risk and Technological Culture: Towards a Sociology of Virulence*, Routledge, London

Walsh, W. J., Reible, D. D., Hass, C. N., Pardue, J. N. and Bowers, T. S. (2006) 'An evaluation of chemical contamination in the aftermath of Hurricane Katrina', *Daily Environment Report*, vol 6, pp1–13

Woodhouse, E. J., Hess, D. J., Breyman, S. and Martin, B. (2002) 'Science studies and activism: Possibilities and problems for reconstructivist agendas', *Social Studies of Science*, vol 32, pp297–319

2
Organizational Culture and the Katrina Response in Louisiana

Rachel A. Dowty, Peter May, William Wallace and Colin Beech

Introduction

Assessments of the federal response to Hurricane Katrina find the Federal Emergency Management Agency's (FEMA) response a failure, the US Coast Guard's relatively successful, and the response of other organizations variable. What explains this variation in organizational performance? The conventional wisdom is that the extraordinary nature of the event in terms of scale and regional impacts created an impossible task for response organizations at all levels of government. But these explanations fail to explain differences between organizations in terms of their performance. We argue that, the role of organizational culture in shaping responses to Hurricane Katrina needs to be considered in order to help explain the differences in organizational performance. This lens provides important insights about the response capabilities of various organizations and the way that leadership from these organizations handled criticism of their performance.

A strong case has been made by scholars who study public agencies that organizational culture serves as an important glue that guides agency commitments (Khademian, 2002, p3) and ways of thinking about central tasks (Wilson, 1989, pp91–93). Various studies of agency shortfalls highlight the role of culture in agency performance (see Vaughan, 1996; Daniels and Clark Daniels, 2000; Romzek and Ingraham, 2000; Waugh and Streib, 2006). Given the seeming importance of culture in shaping agency performance, it seems logical to examine how organizational culture affected the differing organizational responses in the aftermath of Hurricane Katrina. Although a variety of ways of conceptualizing organizational culture have been suggested (see Ouchi and Wilkins, 1985), most scholars end up resorting to fairly inductive depictions that capture the feel of an organization or parts of it. This leads to the divergent depictions of 'command and control', 'safety', 'collaborative' and other cultures noted in the various studies of public agency shortfalls.

In overcoming this conceptual deficiency, we find it useful to work with the concept of 'cultural bias' and classification of four distinctive cultures developed by anthropologist Mary Douglas, and extended in collaboration with political scientist Aaron Wildavsky (Douglas, 1999; Douglas and Wildavsky, 1982). This conceptualization is of particular relevance to our investigation because it explicitly considers the role of risk in shaping organizational performance. To our knowledge, only Christopher Hood (1998) has applied this perspective to the study of what Hood labels 'public-management disasters'. We further develop this theory into a more rigorously delineated methodology that makes it broadly applicable to studies of organizations responding to crises.

We first present the cultural theory of Douglas and her collaborators as it relates to cultural biases of organizations and corresponding responses to risk. We follow this with an explanation of the classification criteria for different preferences and rationales organizations exhibited when responding to crisis tasks that arose from Hurricane Katrina in Louisiana. Response strategies enacted by organizational representatives help to identify the dominant cultural biases evident for the White House, FEMA, the US Coast Guard and emergent response groups. This approach provides practical ways to understand different rationalities that guided organizational responses to the hurricane's aftermath and the justifications that organizational leaders used in responding to criticisms of their performance.

Cultural biases

Everyday parlance generally employs the concept of organizational culture as 'a way of doing business'. In this sense its usage refers to how different organizations go about resolving tasks at hand in different ways (see Ouchi and Wilkins, 1985, p58; Wilson, 1989, p91). In developing this, we relate Douglas' concept of 'cultural bias' to risk and organizations. Douglas defines a 'cultural bias' as 'a steady preference for one or another set of institutional forms and consequently a commitment to the kinds of knowledge that go with it.' (1999, p411). Douglas (1978) developed a classification of cultural biases employing a GRID-GROUP taxonomy that has been elaborated upon in later work among others by Thompson et al (1990). Figure 2.1 presents the typology. An organization's dominant cultural bias characterizes ways that rules structure internal and external interactions (GRID dimension), and ways interactions influence internal and external group ties (GROUP dimension). These two dimensions consistently delineate each of four possible cultural biases: fatalist, hierarchist, individualist and egalitarian.

Douglas' theory assumes dynamic not static cultures, in that all four cultures are always present as competing rationalities for how to make decisions in the face of ever-changing contexts and situations. Thus when an organization, group or individual is said to have a dominant cultural bias, it

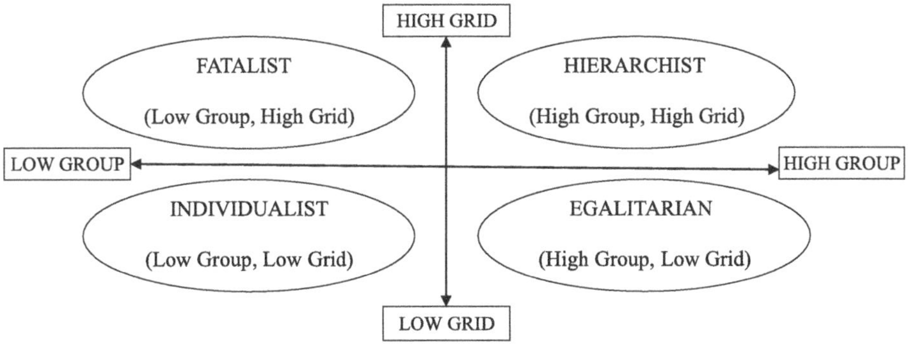

Source: Adapted from Douglas, M. (1978)

Figure 2.1 *Douglas' Typology of Cultural Biases*

does not mean that the other three cultural biases are absent. Instead, it means that the identified cultural bias dominates the other three possible cultural biases given the context(s) of interacting participants. All four cultural biases, and their associated definitions of what constitutes rational decision making, are therefore always present and defining themselves against the others.

Cultural biases, as sets of preferences, have sometimes been referred to by their developers as world views (Douglas and Wildavsky, 1982; Thompson et al, 1990). The biases establish preferences for different types of organizational life, which sociologists think of as ways of organizing shared relationships (see Ouchi and Wilkins, 1985). These include a rigidly regulated and close-knit group with tight boundaries to outsiders (hierarchist), fewer rules and loose group ties (individualist), rigid rules but loose group boundaries (fatalist), or less rigid rules and tightly knit group boundaries (egalitarian). At any given time, an individual or organization exhibits a dominant cultural bias, but ongoing competition among the four biases makes organizational culture dynamic.

Cultural biases and disaster response

Determination and assessment of risk play important roles in the dynamics of cultural biases. Douglas and Wildavsky (1982) argue that each cultural type has its own 'risk portfolio' that leads to differences in alarm, fear and risk reactions. Christopher Hood (1998) employs this perspective in applying the theory to scandals and disasters. He argues that a 'cultural-theory perspective suggests that different sorts of failure are not likely to be randomly distributed around the world of organizations either. Each major organizational way of life is likely to have its own inbuilt Achilles' Heel or characteristic path to collapse' (Hood, 1998, p27). As such, the source of what Hood labels 'public-management disasters' and the response to them is part and parcel of different organizations' cultural biases.

Table 2.1 *Cultural biases and disaster responses*

	Cultural biases			
Responses	*Fatalist*	*Individualist*	*Hierarchist*	*Egalitarian*
Blame (Hood, 1998; Thompson et al, 1990)	The fickle finger of fate; world does things to us.	Faulty incentive structures. Competitive system remains blameless, attribute personal failure to bad luck and/or personal incompetence.	Poor compliance with established procedures, lack of professional expertise. Cannot blame collective system, blame shifted to deviants who don't know their place.	Abuse of power by top-level leaders, system corruption. Blame the collective or the system, solidarity by portraying external symbols as monstrous.
Crisis response (Hood, 1998)	Minimal anticipation, at most ad hoc response after the event.	Market-like mechanisms, competitions and leagues, information to support choice.	More expertise, tighter procedures, greater managerial 'grip'.	Participation, communitarians, whistle-blowing.
Management emphases (Thompson et al, 1990)	Manage neither needs nor resources. Needs and resources are defined by someone else, so copes with that which it has no control over.	Manage both needs and resources. Competitive individualism to manage both upward to the limit of entrepreneurial skill.	Manage resources but not needs. Differential maintenance of resources according to hierarchically patterned levels of needs.	Manage needs but not resources – resources as fixed raw materials that require decreasing needs to equitably share resources.
Typical credo	'I'm not even supposed to be here today.'	'Every man for himself.'	'All for one and one for all.'	'A world in ourselves and in each other.'

Source: Authors, plus adaptation of concepts as indicated from Thompson et al (1990) and Hood (1998)

Table 2.1 presents what we think the different cultural biases suggest about responses to disasters. This draws from Douglas and her collaborators' (primarily Thompson et al, 1990) and Hood's (1998) depiction of cultural biases in organizational responses to disasters. Each cultural type exhibits unique tendencies in its attribution of blame for a disaster, crisis management and disaster response. We also add a 'credo' that captures the organizational perspective of each cultural bias.

Fatalists blame the hand of fate, have ad hoc management and crisis responses, and wonder how they get caught off guard by disasters. Hierarchists blame breakdowns in established procedures and authority, seeking to manage and respond to disasters by tightening standard operating procedures. Individualists blame personal incompetence and/or bad luck for failed management and crisis responses, and emphasize use of market-like mechanisms and rating systems to maximize available entrepreneurial skills. Egalitarian groups blame abuse of power by top-level leaders, and emphasize the need for community participation in crisis response and management to equitably distribute limited resources.

Approach and data

We apply this theorizing in order to understand differences in patterns of blame, crisis responses and management emphases in responding to Hurricane Katrina for FEMA, the US Coast Guard, the White House and emergent groups responding to crisis situations precipitated by Hurricane Katrina in Louisiana. We did this through systematically identifying quotes that explicitly lay blame, described a crisis response, and/or characterized a management emphasis on needs and resources. Transcripts from teleconferences that took place between 28 August and 4 September 2005 at the Baton Rouge Emergency Operations Centre were analysed. The initial analysis of the transcripts assigned each relevant communication and organization to one of the four basic cultural forms. Passages from Congressional testimony of Coast Guard and FEMA personnel, governmental reports, interviews and source-verified newspaper articles were also categorized into one of the four basic cultural forms.

After identifying source materials that explicitly lay blame, stated crisis response strategies and/or actions taken, and explicated management emphases, we identified GRID and GROUP characteristics of each relevant passage. For example, William Lokey, FEMA Federal Coordinating Officer of Louisiana Hurricane Katrina Response and Recovery, testified that:

> *The breakdown of the levee system surrounding New Orleans is the major event that made this disaster truly catastrophic. Without that, it would have been a terrible hurricane, but not the catastrophe it became. Hundreds of FEMA Urban Search and Rescue personnel took an active part in the largest water rescue operation in history, even though water rescue was not part of their mission.* (Lokey, 2005, p4)

Lokey's reference to FEMA's Urban Search and Rescue's mission suggests strict rule structure (high GRID), but that those rules did not apply in the crisis situations in which teams found themselves. This quote does not clearly indicate a GROUP dimension, so no direct conclusions concerning GROUP could be drawn from this quote.

We derived the GROUP dimension by using patterns of blame, crisis response and management emphasis for each of the cultural biases. This quote highlights the role that Lokey assigns the fickle finger of fate (levee breakdown) to play in FEMA's struggles to keep up with crisis tasks, suggesting a fatalist preference for assigning blame. He also describes FEMA Urban Search and Rescue personnel as untrained for water search and rescue (suggesting the fatalist's credo of 'I'm not even supposed to be here today'), but that they did what they could when they could, also indicating a characteristic fatalist ad hoc crisis response. Lokey's statement also clearly states that FEMA was overwhelmed and had no control of the crisis situations fate dealt them, such that they could not effectively manage needs or resources (fatalist management emphasis). Thus, this quote from Lokey characterizes a low GROUP dimension, and a dominant fatalist organizational representative's world view when reflecting on crisis situations that faced FEMA personnel.

Quotes taken from governmental assessments and reports of the response, source-verified newspaper articles, interviews and Congressional testimony provided more reflection from organizational representatives about an ongoing or past response, and included more instances of blame than were explicit during the teleconferences. Emergent groups did not participate in the teleconferences. However, conversations regarding unutilized volunteers and 'security concerns' are traced from these initial reports to later confirmation of their activities though source-verified newspaper reports and government documents. Emergent groups considered for this initial study include citizens and other impromptu groups formed by on-the-scene individuals from various agencies and occupations.

The organization's dominant cultural bias was identified as the one most frequently exhibited and/or described by the patterns of blame, crisis responses and management emphases of the organization's representatives throughout the days for which teleconference transcripts were examined. Several steps were involved in identifying the most frequently exhibited bias for each organization for each teleconference. First, all selected quotes for each day's teleconference were organized according to organization (according to which organizational representative was doing the talking). Second, we identified patterns of blame, crisis response and management emphasis, as done with the aforementioned quote from William Lokey's Congressional testimony. Through these characterizations, we identified the cultural bias exemplified by each quote (such as characterizing Lokey's quote as depicting a fatalist cultural bias' blame, crisis response and management emphasis).

Once we had characterized each quote from each day's teleconference as fatalist, hierarchist, egalitarian or individualist, the number of fatalist quotes from each organization's representatives during each teleconference was counted. The number of individualist quotes, egalitarian quotes and hierarchist quotes that spoke of blame, crisis response and/or management emphasis was also counted for each day's teleconference. This provided a measure of the extent to which each organization displayed each of the four cultural biases for each day's teleconference. These counts enabled us to discern the number of times each organization displayed characteristics of low grid, high grid, low group and high group during each day's teleconference. The relative rise and fall in the number of times organizations displayed high GRID, low GRID, high GROUP and low GROUP are depicted in Figure 2.2 (page 36).

Dominant organizational cultural biases

Quotes from Baton Rouge Emergency Response Centre teleconference transcripts provided real-time response documentation of interactions among responding agency representatives between 28 August and 4 September 2005 (Katrina made landfall in Louisiana on 29 August). The analysis of teleconference transcripts indicated changes in how rules were followed and how group boundaries were set according to blame, crisis response and management emphases, thereby allowing recognition of rising and falling levels of GRID and GROUP. For example, on 30 August, the participating representative from FEMA Operations Headquarters is quoted during the teleconference as saying, 'We are still acquiring satellite imagery and analysis through the National Geospace Agency, and we are concentrating on maintaining visibility of all life support and life-saving issues in coordination with all ESFs [Emergency Support Functions] here at Headquarters.' (30 Aug., pp55–56) This quote describes problems that FEMA is having with maintaining coordination amongst its own employees, as well as suggesting that their rule structures are playing out such that they can better accomplish goals (visibility) by working through another agency entirely (the National Geospace Agency). Both these strongly suggest low GROUP. The goal they are working toward (visibility) suggests a focus on managing neither needs nor resources, but instead managing the determination of events to later determine what needs and resources should be sought (suggesting an ad hoc response). These are crisis responses and management emphases characteristic of a dominant fatalist organizational culture (high GRID, low GROUP).

Two days later, William Lokey states during the daily teleconference that 'It's very difficult for us to be shopping around the greater United States for places… I know that has some real Stafford Act implication issues, but with the numbers that are growing of what we have to move out of New Orleans, it's just we have nothing anywhere near here to put all these folks.' (1 Sept., p13) This quote more clearly exemplifies representation of a fatalist organization than the quote from William Lokey two days earlier. It not only shows an ad hoc response and tendency to manage neither needs nor resources, but also shows blame aimed toward the workings of (high GRID) rule structures (the Stafford Act) that were impeding an effective ad hoc response. Through such comparisons of daily quotes from representatives of the same organizations (i.e. FEMA representatives, White House representatives, Coast Guard representatives and mention of emergent groups), trends in rising and falling levels of GRID and GROUP for each organization were traced and charted. Clearly, it is impossible to encapsulate all of the different responses. But, we can illustrate how actions adhere to cultural biases.

The graphs in Figure 2.2 show trends in organizational grid and group characteristics exhibited during Baton Rouge Emergency Operations Centre teleconferences over the first week of disaster response in and around the New Orleans area. FEMA begins with a slightly higher GRID and lower GROUP, while over the course of the week this trend continues unabated. This suggests

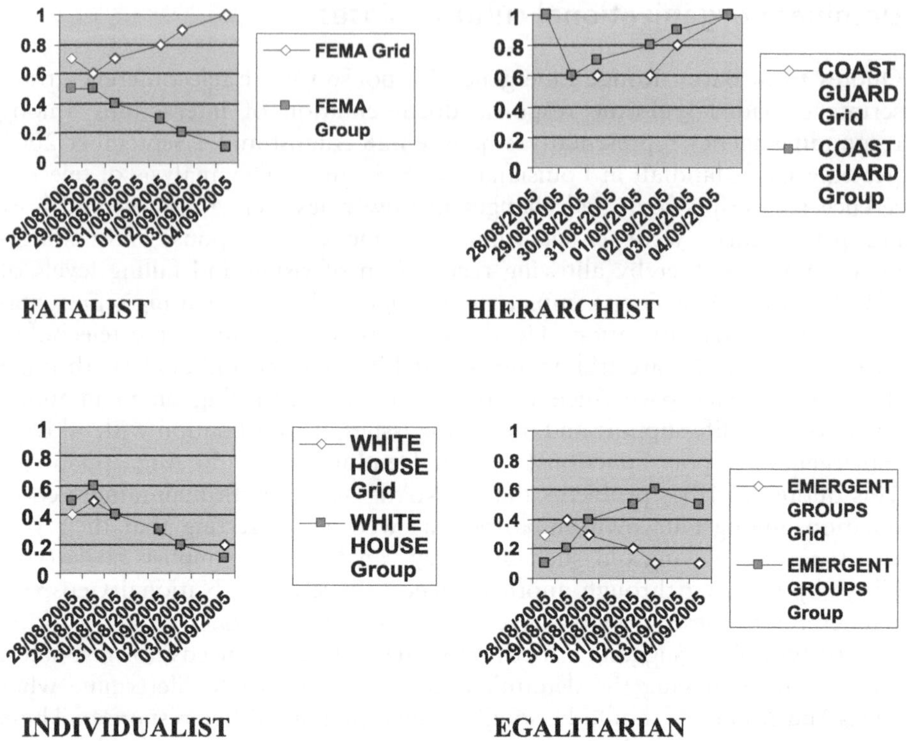

Note: Each graph shows relative increases and decreases in GRID and GROUP over the first week of disaster response

Source: Compiled by authors

Figure 2.2 *Cultural biases evident in Katrina emergency operations*

FEMA exhibited a trend toward a dominant fatalist cultural bias (high GRID, low GROUP). The Coast Guard shows a fluctuating but overall hierarchist (high GRID, high GROUP) cultural bias. The White House had an increasing individualist cultural bias (low GRID, low GROUP) over the first week of disaster response. Emergent groups tended to move toward egalitarian (low GRID, high GROUP) organizational arrangements.

The high GRID of the Coast Guard and FEMA had substantially different consequences for the way each approached their tasks because of their differences in GROUP. For the Coast Guard, a coupling of high GRID and high GROUP enabled organizational sub-groups to integrate more readily and securely improvise rules of action in the absence of communication with their regular (non-disaster) chain of command. In other words, the high GROUP of the Coast Guard enabled organizational members to improvise rule and group structures necessary to continue resolving tasks under extraordinary conditions. FEMA, on the other hand, had just as strict a hierarchical rule structure as the Coast Guard, but no GROUP to facilitate coordination within or between organizational hierarchies. Consequently, FEMA's fatalist cultural bias rendered rule structures a hindrance to resolving disaster tasks, because the organization lacked group integration. Rules became obstacles to 'push' away, instead of empowering group improvisation as evidenced by overall Coast Guard performance. The most obvious consequence of FEMA's fatalism when faced with tasks that also faced the Coast Guard was when a Coast Guard representative (Adm. Thad Allen) replaced a FEMA representative (Michael Brown) as Principal Federal Official (PFO) to oversee the response to Hurricane Katrina in Louisiana.

Fatalist FEMA, with its low GROUP, preferred to interact more with White House staff than with Department of Homeland Security (DHS) officials, as evidenced by Michael Brown's calls to White House staff instead of efforts to communicate with Michael Chertoff or others that follow command structures. However, Brown's communications to White House staff resulted in few tasks being resolved by FEMA, and instead resulted in FEMA handing tasks off to the Coast Guard and using FEMA's command structure to find where to place blame. It is significant that Michael Brown identified himself as a scapegoat during the first few days of the response, before failure had even been ascertained. It indicates that the organization's top people regarded FEMA's response a failure before the organization's response was even underway.

The White House, sharing FEMA's low GROUP but not FEMA's high GRID, took the opportunity Brown handed them, and shifted blame when and where possible. The White House's 'every man for himself', low GRID, low GROUP individualist cultural bias facilitated within-organization avoidance of blame and responsibility by shifting it onto state governments and fatalist FEMA. Instead of focusing on what actions should be taken in teleconference calls, White House representatives avoided action, preferring to focus on giving praise and words of encouragement. They avoided participation in conversations that required strong rule structures and tightly knit groups, and through praise reinforced the belief that other organizations with rules and groups were better equipped and able to do what needed to be done.

Groups that emerged on-the-scene were necessarily tightly knit and in close proximity (in rescue boats, for example), and the Coast Guard's high GROUP fit in well and facilitated ad hoc group formation in search and rescue operations. However, the absence of rule structures held within emergent groups failed to impart a structure that enabled continued communication once groups were physically separated (in evacuations or in response assignments). This resulted in a process wherein egalitarian groups formed when people hung together for mutual support and survival, then dissolved because nothing but their integration under disaster conditions held them together to begin with. Rule structures were the things getting in the way, the reason they were forming their ad hoc groups in the first place, thereby giving a clear example of the egalitarian tendency to shift blame to authority figures and system corruption. These emergent egalitarian organizations managed needs, but used their resources too quickly in managing those needs to achieve many sustainable organizational structures.

Emergent groups tended to move toward egalitarian (low GRID, low GROUP) organizational arrangements. The fact that there are changes for emergent groups from a slightly higher GRID and lower GROUP (suggesting a fatalist bias) to a clearly low GRID, high GROUP (egalitarian) bias suggests that groups of people who were not tightly knit became increasingly so as the first week of disaster response wore on. It may also suggest that rules guiding the people who formed these groups were not working as they expected, leading to a decrease in regulation by rules to an increased level of group solidarity.

Cultural biases and organizational responses

The flavour of the differences in cultural biases comes out more strongly when considering comments that were coded from the teleconferences.

FEMA – Fatalist bias

After FEMA was absorbed into the DHS in 2003, formalized rule structures remained, but collective organizational solidarity was lost. As documented by the FEMA Office of Inspector General (2006), changes in task assignments and inconsistent expectations from above undermined the group solidarity. This suggests that, between the time that FEMA was incorporated into the DHS and when the Katrina disaster struck, FEMA experienced a high GRID (formalized rule structures) and low GROUP (low collective solidarity), or fatalist, cultural bias. Fatalists tend toward an apathetic or ambivalent way of life, where fate receives the credit or blame for new developments. 'I'm not even supposed to be here today' captures a characteristic fatalist view of events.

This bias is evident from the teleconferences. During teleconferences, FEMA Director Michael Brown began on 28 August by emphasizing the need for everyone to 'push the envelope' as far as possible, without going over 'the edge'. One can think of the envelope as the rules and structure of grid-group that need to be abandoned. In Brown's words:

> *I want to emphasize what I said yesterday, get to the edge of that envelope. And, in fact, if you feel like you [missing], go ahead and do it. I'll figure out some way to justify it, some way [missing] tell Congress or whoever else it is that wants to yell at me, just let them yell at me, [missing] not to worry about – in fact, I don't want any of these processes in our way.* (28 Aug., p37)

Brown talks of his concerns, but not how to address them; only that he wants someone else to address them.

A sense of being cut loose from the DHS hierarchy is evident. Patrick Rhode, FEMA Deputy Director, comments at one point: 'I'm not certain if the Department of Homeland Security Secretary Chertoff is with us or not at this point. I would just like to ask if he's there?' (Jackson answers, No), (30 Aug., pp57–58). William Lokey [(Chief of Operations Branch in the Response Division at FEMA, Federal Coordinating Officer (FCO) for the response to Hurricane Katrina in Louisiana until Admiral Thad Allen was appointed FCO for all of the Gulf Coast operations and Lokey became the Deputy FCO for Louisiana)] goes over New Orleans' conditions and says priority is search and rescue (as does everyone else at this point) (29 Aug.). However, Lokey's passivity grows over the following days.

US Coast Guard – Hierarchical bias

As a branch of the military, the Coast Guard constitutes a stereotypical hierarchical organization, bound together through both close 'unit' (high GROUP) ties and a clear formalized rule structure (high GRID). These characterize a hierarchical cultural bias for which 'all for one and one for all' is the call to arms. In contrast to FEMA, the US Coast Guard maintained its identity as a semi-autonomous unit when it became part of the DHS. And unlike FEMA, no efforts were made to reassign Coast Guard responsibilities. As such the strong grid–group relationships remained. As we describe below, this does not mean that the US Coast Guard is a monolithic entity for which subordinates were waiting for orders from above before proceeding. To the contrary, the decentralized authority granted within the Coast Guard and the strong group ties contributed to a more successful response effort.

Consistent with this, the Coast Guard is clearly thought of in the teleconferences as a 'can do' organization. DHS Secretary Chertoff and FEMA Director Brown, when referring to the Coast Guard, always do so in a context of suggesting that Coast Guard personnel are the ones able to assist and get things done when other organizations fail to complete a task. Chertoff tells Mike Brown 'if there's anything that you need from the Coast Guard or any other components that you're not getting, please let us know. We'll do that for you. OK' (28 Aug., p38).

The Coast Guard's differential maintenance of resources according to hierarchically patterned needs also shows in discussions of where, for example, airboats and helicopters (1 Sept., pp19–20) should be delivered and why. The people who are the first priority are Coast Guard personnel: discussions by Coast Guard personnel are concerned with getting the needed resources to Coast Guard personnel so that the stranded civilians can get access to these resources through the Coast Guard. However, the flexibility in those channels via the incident command structure is also apparent in the growing urgency surrounding resource management as the days passed. As Col. Jeff Smith of Louisiana stated on 1 September:

> *You know, I think at this point there's a lot of people worrying about who's in charge. We've still got people that need rescuing. We've got people that need food, water, ice. We need buses. Get it down here and we'll worry about who's in charge. Just get us the assets, please. ... One other comment.*

> *Our Governor has said that we're going to follow the National Response Plan, work within the incident command structure, and we're prepared to do that.* (Smith, 1 Sept., p18)

The White House – Individualist bias

Various players in the White House, including the President, his direct advisors, and various cabinet members were drawn into the Katrina disaster response. While these individuals do not constitute an organization per se, they constitute a loose knit group coordinated by the White House. As shown by the data of Figure 2.2, their actions reflect an individualistic (low GRID, low GROUP) cultural bias. Individualists embrace competition, a belief in personal competence to resolve tasks and minimization of rules. Indeed, President Bush and DHS Secretary Chertoff exhibited ignorance of several rules governing disaster declarations, but nevertheless continued with business as usual for the first few days while the Katrina disaster gained momentum in New Orleans. 'Every man for himself' may be considered an individualistic credo.

White House representatives who participated in teleconferences keep track of where Bush and Chertoff were, and reassured others that things are being done, while thanking people. President Bush, in particular, and Secretary Chertoff generally, maintained a hands-off perspective on relief operations – leaving details to subordinates. They evidence a good deal of faith that things would work out, consistent with the embracement of risk by individualists. Solidarity is found in reassuring others of their competence and having them reassure you of yours in turn.

On 4 September, DHS Deputy Secretary Michael Jackson focuses his questions on getting information concerning numbers of people needing rescuing, and says, 'Any idea about numbers of people – I mean, even a wild guess. In the downtown area, are there still some considerable number of people in the outlying parts of the city? Do we have any sense at all – I'm not trying to hold you to a number, but it's just an order of magnitude' (4 Sept., pp9–10). Maggie Grant, another White House representative, received information through the Exxon refinery in Chalmette, Louisiana (in St. Bernard Parish, bordering New Orleans) that the President of St. Bernard had 'declared some kind of "martial law" and is using his authority to remove key pieces of equipment from the refinery, including front-end loader and crane essential to getting coker units back up and refinery running ... [I] wanted to bring that all to your attention, if someone could maybe check in to see the validity of this question from the Exxon team' (4 Sept., p12). White House personnel use teleconferences to try and ascertain information, not in terms of taking specific actions, but in terms of competitions over needs and resources (e.g. greater numbers of people mean greater priority, Parish President's rumoured acquisition of resources from an oil refinery).

Emergent groups – Egalitarian bias

An emergent group can be as large or small as circumstances dictate, and be composed of any mix of people from anywhere: a policeman, a schoolteacher, a drug dealer, children, ad infinitum. Yet they stick together as a group. These emergent groups have loosely defined internal roles due to lack of a formalized rule structure (low GRID), but 'are perpetually engaged in building up the boundary separating themselves from others' (Thompson et al, 1990, pp239–240). In other words, they fall in the *egalitarian* cultural bias where their organization is strongly incorporated (high GROUP) but has weak structure (low GRID). 'A world in ourselves and in each other' marked the egalitarian cry of radical abolitionists like William Lloyd Garrison.

While emergent groups were not directly represented in teleconference transcripts, they were referred to starting on 2 September. References to volunteers sitting around with nothing to do are included in frustrated remarks by Col. Jeff Smith (2 Sept., pp3–4), who gets a single word reply from Patrick Rhode: 'understood', followed by more rhetoric about how hard everyone is working and how much their work is appreciated. Josh Filler, (now former) Director of the Office of State and Local Government Coordination and Preparedness at the DHS emphasizes that there are many people 'who want to help, but we have encouraged them not to self-deploy to New Orleans or to Louisiana, but to work through the system, but they are saying that their States are not receiving requests for assistance' (2 Sept., p9). Filler was referring here to law enforcement personnel throughout the country. As people (both untrained volunteers wanting to help and deputies, sheriffs, etc.) waited around, they began to form loose groups, some of which self-deployed to New Orleans.

News that self-deployed groups were making their way into New Orleans heightened security concerns that were apparent by 1 September. Response from the government was generally to shift from search and rescue priorities to security priorities, despite the protests of Col. Jeff Smith in teleconference transcripts. For example, on 3 September, Staff Sgt. Paul Hall of the National Guard told *Washington Post* reporters that 'We have to look out for snipers, unfortunately. That's been a big problem down here. They're like terrorists' (Whoriskey and Levine, 2005). However, later reports from the Government Accountability Office and other governmental and private organizations state that 'security concerns' were grossly exaggerated, and coupled with sweeping rumours, significantly slowed already-slow rescue and relief efforts. By redefining the mass of stranded people as victims of terrorists, the organizational management strategies enacted by the White House ever since the formation of the DHS could be enacted.

This shift charged many emergent groups, and seemed to coincide with a heightened sense of solidarity among them. First, as a representative of the State of Texas said, 'We've really created an uncontrolled exodus versus a planned evacuation here.' This caused a tightening of group boundaries (increase in GROUP) among those who were being lined up according to age (which separated the weak elderly and children from their caretakers and families) and medical condition (again, grouping the weak together). It caused a tightening of group boundaries among those wading and floating through the streets of New Orleans. Numerous experiential accounts of these people have and continue to be recorded and published. One notable account was from a member of a racially charged hate group who, upon seeing those he hated helping those he came to help, changed his world view and could no longer accept his reasoning for hate without question, and joined those he had hated in conducting boat rescues.

Secondly, it charged emergent groups by giving them a clear authority against which to set their victimization. This is clearly and repeatedly shown in justifications given by members of emergent groups for their actions. For example, many firefighters from around the country thought they were being assembled by FEMA to assist with crises along the Gulf Coast. They soon learned their role was to disseminate a phone number and fliers as FEMA community-relations workers. Many assembled at a staging area 'peeled off their FEMA-issued shirts and stuffed them in backpacks, saying they refuse to represent the federal agency' (Rosetta, 2005). The slowness and lack of response coordination also prompted solidarity among groups of 'looters'. The depiction of looters as 'terrorists' contradicts footage and experiences of those on-the-scene, some of whom were looting stores alongside police officers and were witnessed being robbed and threatened at gunpoint by National Guardsmen.

Ties that held the emergent groups together began to wane at the end of the first week, although further data analyses are needed to find if this trend continued into the following weeks. Waning group ties would not be surprising, when one considers that obstacles to communication and staying in physically assembled groups grew over the subsequent days and weeks, even for otherwise close-knit families.

Justifications of disaster responses

Justifications are frequently mentioned, but rarely enter into the thick of social science analyses (Boltanski and Thévenot (2006) provide refreshing exceptions to the rule). We demonstrate how cultural biases shaped the justifications that organizational representatives used to focus blame on failed crisis responses and ineffective management emphases. Justifications used by representatives of FEMA, the Coast Guard and the White House reflect the results depicted in Figure 2.2, with fluctuations in GRID and GROUP but overall steady preferences for assigning blame according to fatalist, hierarchist and individualist cultural biases, respectively.

Michael Brown (aka 'Brownie'), former Undersecretary for Emergency Response and Preparedness, former Director of FEMA, and original PFO for the federal response to Hurricane Katrina, blamed a botched federal disaster response on a 'culture clash' that occurred when FEMA was 'folded into' the DHS (Brown, 2006). As a fatalist organizational representative, Michael Brown saw FEMA 'folded into' the DHS, the fickle finger of fate handing down rules (high GRID) to an organization without close ties with those handing the rules down (low GROUP). Cooper and Block (2006, ch 10) document Brown's ongoing apathy and fatalist blame patterns during the following weeks.

DHS chief Michael Chertoff singles out individuals to blame from federal to local levels. Statements not only from Chertoff but other representatives of the White House, including President Bush, explicitly lay blame on Michael Brown, Louisiana Governor Kathleen Blanco and New Orleans Mayor Ray Nagin. One quote that exemplifies individualist blame came from Russ Knocke, spokesperson for Chertoff, who told the *Washington Post* that 'Brown had "all authority" to make decisions and requests, and his "willful insubordination ... was a significant problem" for Chertoff' (Hsu, 2006). As an individualist organization, the White House blamed personal incompetence and lack of sufficient incentive signals to prompt an efficient disaster response.

Representatives of the US Coast Guard maintained characteristic hierarchist patterns of blame. Admiral Thad Allen testified that he believes that FEMA belongs in the DHS (in opposition to Michael Brown's testimony), emphasizing that 'the Coast Guard's relationship with FEMA has been strengthened since the creation of DHS' (Allen, 2006). This exemplifies the hierarchist approach to blame. No blame of the system is allowed for which Allen does not blame the DHS, the organization in which the Coast Guard is situated. As well, the importance of rule structures (high GRID) and strong organizational relationships (high GROUP) are emphasized. In his Congressional testimony (Collins, 2005), Coast Guard Commandant Thomas Collins blamed public misunderstandings of the Coast Guard's organizational mission as affecting perceptions of the agency's performance.

Conclusions

This chapter shows how cultural biases provide different lenses through which various organizations faced with similar disaster-related circumstances end up responding differently and later justify those responses in different ways. This perspective helps explain organizational decision-making in crisis situations and demonstrates the difficulties of improving planning and coordination within and among diverse disaster-response organizations. Different ways of rationalizing actions lie at the heart of all of this. We have shown how organizations that have different dominant cultural biases respond to the same disaster situations with different presumptions and rationales for their decisions. Rational action for one organization's dominant culture may not be deemed rational by another dominant culture during crisis responses. The conflicts that this engenders are usually only detectable after the fact, far too late to overcome shortfalls in disaster response.

The type of analysis of organizational culture undertaken for this research opens up avenues for assessing how working relationships between disaster agencies might be strengthened for disaster response. This type of analysis can lead to new understandings of the interplay of different agencies as they plan for and respond to disasters. For example, Weichselgartner and Brévière (Chapter 8) review the historical and cultural contexts of disaster response in Germany. They note a clear shift has taken place over the years from the Catholic Church spearheading disaster response to reliance on a network of governmental laws and regulations. Because both the Church and the government operate fundamentally as hierarchical organizations (high GRID), attention should focus on the issues of group solidarity (the GROUP) and how that affects the ability to form working relationships between the two entities. Finding unexpected similarities among organizations responding to disasters may prove just as, if not more, useful when planning organizational collaborations.

The cultural biases of different organizations should be considered when designing drills and training exercises for disaster response. For example, FEMA and the Coast Guard's cultural biases could have been taken into account during 'Hurricane Pam' training exercises that took place in Louisiana in the summer of 2004. This was a series of training exercises for preparing 'state and federal agencies to help officials develop joint response plans for a catastrophic hurricane in Louisiana' (FEMA, 2004). If the different organizational cultures had been considered as part of the training, the leaders of the two organizations would have had a better means of interpreting what was happening as the agencies responded to Hurricane Katrina. Failures in communication would have been seen less as not wanting to work together and more as problems of different ways of interpreting what was going on. More generally, assessments of organizational cultural biases with respect to risk may better predict how an organization can improvise and coordinate within its own ranks and with other organizations.

None of this is within the realm of normal organizational development and planning. As a consequence, leaders need to be attentive to ways to build in such 'organizational thinking' and the opportunities it presents for learning. Training people already poised in positions within organizations to understand the ways that organizational cultural biases affect decisions within an organization can help. The use of computer-based simulations, much like existing table-top disaster exercises, provides another means for raising awareness of the cultural issues. Regardless of these particular suggestions, the basic lesson of this contribution concerns the importance of expanding the understanding of the ways that organizational cultural biases play out in disasters.

Acknowledgements

Financial support for this research is provided by the National Science Foundation (NSF), Small Grants for Exploratory Research (SGER) grant number SES-0554938, and Human and Social Dynamics grant number SES-0623907 made to Rensselaer Polytechnic Institute under Principal Investigator William Wallace. NSF is not responsible for data, analyses or claims made by the authors.

References

Allen, T. W. (2006) 'Testimony of United States Coast Guard Admiral Thad W. Allen, Commandment on FEMA's Placement in the Federal Government', *United States Senate Committee on Homeland Security and Governmental Affairs*, 8 June, Federal News Service, Washington DC

Boltanski, L. and Thévenot, L. (2006) *On Justification: Economies of Worth*, translated by Catherine Porter, Princeton University Press, Princeton, NJ

Brown, M. (2006) 'Testimony of Michael Brown, Former Undersecretary for Emergency Response and Preparedness', *United States Senate Homeland Security and Governmental Affairs Committee*, 10 February, Federal News Service, Washington DC

Collins, T. (2005) 'Testimony of United States Coast Guard Commandant Thomas H. Collins', *United States House Homeland Security Subcommittee on Science and Technology and the Armed Services Subcommittee on Terrorism, Unconventional Threats and Capabilities*, 9 November, Federal News Service, Washington DC

Cooper, C. and Block, R. (2006) *Disaster: Hurricane Katrina and the Failure of Homeland Security*, Henry Holt and Company, New York

Daniels, R.S. and Clark-Daniels, C. (2000) 'Transforming government: The renewal and revitalization of the Federal Emergency Management Agency', Report to the Price Waterhouse Coopers Endowment for the Business of Government, The Endowment, Arlington, VA

Douglas, M. (1978) 'Cultural bias', Royal Anthropological Institute of Great Britain and Ireland, Occasional Paper no 35, 59pp

Douglas, M. (1999) 'Four cultures: The evolution of a parsimonious model', *GeoJournal*, vol 47, pp411–415

Douglas, M. and Wildavsky, A. (1982) *Risk and Culture: An Essay on the Selection of Technological and Environmental Dangers*, University of California Press, Berkeley, CA

Federal Emergency Management Agency (FEMA) (2004) 'Hurricane Pam exercise concludes', FEMA Press Release R6-04-093, 23 July 2004 www.fema.gov/news/newsrelease.fema?id=13051 accessed 20 August 2009

Hood, C. (1998) *The Art of the State: Culture, Rhetoric, and Public Management*, Oxford University Press, Oxford

Hsu, S. S. (2006) 'Katrina report spreads blame: Homeland Security, Chertoff singled out', *The Washington Post*, 12 February, pA01

Khademian, A. M. (2002) *Working With Culture: How the Job Gets Done in Public Program*, CQ Press, Washington DC

Lokey, W. (2005) 'Testimony of William Lokey, Federal Coordinating Officer for the Louisiana Hurricane Katrina Response and Recovery', *United States House Select Committee to Investigate the Preparation for and Response to Hurricane Katrina*, 14 December, Federal News Service, Washington DC

Office of Inspector General, United States Department of Homeland Security (2006) 'A Performance Review of FEMA's Disaster Management Activities in Response to Hurricane Katrina', Office of Inspections and Special Reviews, OIG-06-32, 109th Cong., 2nd sess., Washington DC, p24

Ouchi, W. G. and Wilkins, A. L. (1985) 'Organizational culture', *Annual Review of Sociology*, vol 11, pp457–483

Romzek, B. S. and Ingraham, P. W. (2000) 'Cross pressures of accountability: Initiative, command, and failure in the Ron Brown Plane Crash', *Public Administration Review*, vol 60, no 3, pp240–253

Rosetta, L. (2005) 'Scores fear escape means losing it all', *The Salt Lake Tribune*, 6 September, pA1

Thompson, M., Ellis, R. and Wildavsky, A. (1990) *Cultural Theory*, Westview Press, Boulder, CO

Vaughan, D. (1996) *The Challenger Launch Decision*, University of Chicago Press, Chicago, IL

Waugh, W. L. Jr. and Streib, G. (2006) 'Collaboration and leadership for effective emergency management', *Public Administration Review*, vol 66, no 1, pp131–140
Whoriskey, P. and Levine, S. (2005) 'Guard troops descend on New Orleans', *The Washington Post*, 3 September, pA01
Wilson, J. Q. (1989) *Bureaucracy: What Government Agencies Do and Why They Do It*, Basic Books, New York

3

Disasters as System Accidents: A Socio-ecological Framework

Gary Bowden

Previous accounts of Hurricane Katrina have focused on two disasters: the hurricane and the flood. In contrast, I highlight a third, socially based disaster: the unusual and unexpected amount of civil strife that emerged following Katrina. Although structured as an explanation of that civil strife, the wider purpose of this chapter is to develop a theoretical perspective that fuses Klinenberg's (1999) socio-ecological perspective with Perrow's (1999) concept of a system accident in order to understand disasters as an emergent characteristic of socio-ecological systems involving interactions among three different systems (natural, technological, social) each of which operate across a variety of spatial-temporal scales (from micro to macro).

Several important points flow from this perspective. First, it treats individual events as the emergent result of a complex set of interacting processes. By drawing attention to these processes, it is possible to build scenarios and forecasts that can effectively anticipate future events. While not a particularly novel idea, existing scenarios focus almost exclusively on natural and technological factors. The novelty of the present analysis, as in Chapter 2 by Dowty et al, lies in the role attributed to the socio-cultural context, a topic recognized as having profound effects on theory construction (Abbott, 2005), and largely overlooked in the disaster studies literature (Alexander, 2006). Unlike Dowty et al, who use the concept of organizational culture to understand differential patterns of *disaster response*, this chapter emphasizes the significance of culture for understanding the unique nature of a particular disaster and the contextual *causes* of that uniqueness.

Second, the perspective highlights the wide number of processes underlying any given phenomenon and, hence, the need for organizing frameworks that emphasize intellectual breadth. Human knowledge has successfully advanced largely through specialization; by breaking reality down into manageable chunks and analysing them in detail. As Frickel and Vincent (Chapter 1) have noted, the production of knowledge is an organized social process that often produces 'knowledge gaps'. Moreover, the emphasis on depth can create blinders – that is, situations where relevant knowledge actually exists but is not perceived as relevant to the issue by the extant interpretive schema. For example, attempts to maximize crop yield in the 1950s led agricultural chemists to advocate the widespread use of pesticides. Only through the application of a broader perspective, one that required attending not only to crop yield but also to the consequences of the pesticides on other aspects of the ecosystem, could Rachel Carson write *Silent Spring*. There is currently a serious need for analytical frameworks that effectively integrate diverse bodies of specialized knowledge. The conjunction of system accidents and social-ecological analysis presented in this chapter exemplifies one such framework. I begin by throwing light on a phenomenon overlooked by standard accounts of Katrina and then develop an explanatory framework designed to make sense of the overlooked findings.

What needs explanation?

Natural disasters are fundamentally a social phenomenon; their severity is measured in human terms, lives lost or dollars worth of devastation, rather than scientific terms. Thus, no one called the 7.0 magnitude earthquake that hit the Mojave Desert region of California on 17 October 1999 a disaster because the human impact – damage to a few roads, knocking items off shelves, and so on – was slight. In contrast, the 12 January 2010 quake of the same magnitude that devastated Haiti is universally labelled a disaster and will be remembered for decades. But the human impact of a disaster extends beyond actuarial statistics and measures of economic loss. In the case of Katrina, there is also the role of the disaster in exacerbating social conflict. While the bulk of this chapter focuses on the tensions within New Orleans immediately following the hurricane, these tensions should not be viewed in isolation. They are only the most immediate manifestation of a dynamic that has ongoing consequences evident in, for example, the number of blacks who have been forced to relocate outside New Orleans and the ongoing controversies over how to rebuild the city (Barrios, Chapter 6). In other words, the conflict discussed in this chapter should not be viewed as an isolated phenomenon but, rather, as an emergent manifestation of tensions that existed prior to Katrina, were exacerbated by the disaster and continue to manifest themselves post-Katrina. With that in mind, I begin with a discussion of the typical impact of disasters on social solidarity and then turn to a description of how New Orleans differed, a difference explainable by social context.

Solnit (2009) is the latest in a long tradition documenting the social cohesion, at both interpersonal and community levels, typical of disasters (e.g. Fritz, 1961; Barton, 1969; Drabek, 1986). Faced with a common trauma, individuals put aside their social differences in order to provide material and emotional support for victims. Thus, as was seen in New York following 9/11, one typically finds both self-organizing helping behaviour (e.g. the organization of boats to get people off Manhattan) and indicators of an increased sense of community (e.g. a drop in the crime rate).

While responses of this type, as would be expected and documented by Solnit (2009) and others, predominated in New Orleans in the wake of Katrina, media coverage focused attention on another social dynamic, a descent into lawlessness characterized by looting and violence, and general social chaos (Brinkley, 2006, pp502–509). In the wake of such a large-scale calamity, a substantial amount of 'looting' can be expected as individuals search for food, water and other necessities. I do not count such behaviour as evidence of social conflict. However, the events in New Orleans went substantially beyond the unauthorized acquisition of necessities as, for example, televisions and other non-essential items were taken from stores, only to be abandoned a few blocks away.

Significantly, the atypical behaviour involved both citizens and local officials. Members of the New Orleans Police Department, for example, were criticized, among other things, for: (a) abandoning their jobs; (b) sloughing off (e.g. having large numbers of officers deployed at unnecessary checkpoints); (c) driving through crowds or leaving crisis situations without offering help; and, most surprisingly, (d) using the disaster as an opportunity to steal a number of cars from a Cadillac dealer (United States Congress, 2006, pp240–265).

Many of the most egregious acts attributed to the local citizenry (e.g. the suggestion that individuals had been murdered at the Superdome) turned out to be false (Brinkley, 2006, pp366, 392). However, it is significant to note that these rumours were repeated as fact by high-ranking public officials. In other words, even if they didn't happen, they were taken by individuals with intimate knowledge of the city as *plausible events*. Thus, we are confronted not only with the atypicality of the events in New Orleans but, also, with the existence of a plausibility structure within the local elite that rendered such actions as conceivable.

Consider two specific and symbolically rich examples of the atypical behaviour. The first, widely known as the Gretna bridge incident, involved a group of several hundred exhausted hurricane victims (a mix of tourists, parents with babies and the elderly – some of which were in wheelchairs) estimated at 95 per cent black who, following instructions from an official, walked several miles to a bridge linking New Orleans with Gretna. They were met by Gretna police who, under direct orders from the mayor, forcibly prevented them from crossing the bridge. The evacuees were told 'We will not allow Gretna to become another Superdome', shots were fired over their heads, and a helicopter was used to buzz them and drive them back into New Orleans (Brinkley, 2006, pp469–472). Second, looting was often accompanied by the 'big dump', that is, defecating in cash boxes, deep fryers, on bar-tops, in store aisles, and so on. As one of the store owners noted, 'They were making a statement, shit on you' (Brinkley, 2006, p276).

These examples illustrate the pollution behaviours discussed by Mary Douglas (1966). In the first instance we see coercive force used to maintain purity – to prevent the spread of disorder associated with New Orleans (and particularly blacks) into Gretna. In the second instance we see the ritual pollution of the property of the dominant group by individuals who feel abused and ignored. Pollution, in this case, is used as a mechanism of personal empowerment. These, and other similar incidents, indicate the ongoing importance of race and class during the disaster. This is substantially different from the typical community response to disaster. How do we account for this?

A Congressional investigation emphasized the following factors: (1) a lack of basic necessities (leading to looting in order to obtain food, water, medical supplies, etc.); (2) overcrowding and uncertainty about evacuation (resulting in frayed nerves and desperate individuals); and (3) the 'collapse or absence' of law enforcement (United States Congress, 2006, pp240–260). This account is interesting not so much for what it includes as for what it excludes: any mention of race or class. The account focuses on stresses that would affect anyone placed in the situation, irrespective of race or class standing, and the presumed need for social control in such situations. In short, the Congressional report treats social chaos as a natural and expected state of affairs for people under stress. Without coercive authority to keep local citizens in check, things fell apart until outside coercive forces could bring them back under control.

This account fails on two grounds. First, it is inconsistent with our understanding of typical community reaction during a natural disaster. Second, it ignores the obvious relevance of racial and class tensions in New Orleans as illustrated above. But, it is equally simplistic to suggest that racial and class divisions alone are responsible. New Orleans has a long history of being hit by hurricanes. And there is no evidence that those previous incidents dramatically exacerbated existing social tensions. Thus, we need an account that locates the events in both social and historical context. It is to such an account that we now turn.

How do we explain it?

This section links two theoretical frameworks: Perrow's analysis of system accidents and Klienenburg's social-ecological approach. From Perrow (1999) comes the distinction between component failure accidents and system accidents. Component failure accidents involve one or more component failures (parts, units, subsystems) that are linked in an *anticipated* manner. System accidents, by contrast, involve the *unanticipated interaction* of multiple failures. The multiple failures interact in unforeseen ways, frequently creating situations that are strange, unexpected and incomprehensible to the people who experience them. Previous discussions of Katrina have treated the events in New Orleans as a component failure accident. The possibility that a hurricane could breach the levee system was well understood (Congleton, 2006, p27) and, indeed, such a scenario was ranked as the most catastrophic potential natural disaster for the United States by the Federal Emergency Management Agency (FEMA). This view, however, brackets out the social significance of the disaster. It wasn't simply the scale of devastation and the ineptitude of the response that outraged the American public. Katrina brought issues of poverty and race back onto the public agenda, at least for a brief moment. To understand why this was the case we need to treat the New Orleans disaster as a system accident resulting from the interaction of three different systems – a natural system (the hurricane), a technological system (the levees and related water control technologies) and a social system (the social dynamics of New Orleans, particularly in relation to race) rather than as a component failure accident involving only natural and technological elements.

Recent studies of disasters have advocated a historically informed approach that goes beyond the immediate and localized circumstances of the problem under investigation. This chapter follows and extends the logic of this type of analysis, as exemplified in Klienenburg's (1999) analysis of the 1995 Chicago heatwave, Ali's (2004) analysis of the *E. coli* outbreak in Ontario and Gall's use of the concept of social vulnerability (Chapter 9). Specifically, the present study employs a socio-ecological perspective to trace the causal links relevant to understanding the unusual level of social chaos in New Orleans at the time of Hurricane Katrina by explicating the processes attached to three systems (natural, technological, social) at three different scales (micro, meso and macro). These scales are meant to analytically highlight different types of processes – those that operate over short periods of time in relatively localized areas (micro) through those that operate over extended periods of time and large areas of space (macro). It will be shown that the well documented social disorder was not simply an anomalous event limited to one point in time and place, but the emergent product of an extended set of processes that evolved over time and through different spatial scales at the social, technological and biophysical levels.

A schematic representation of the account is shown in Table 3.1. Each row represents the processes associated with a particular system. Each column represents the processes attached to a particular spatial-temporal scale. The framework is an analytical structure intended to highlight and draw attention to the diverse range of processes relevant to understanding what happened in New Orleans and does not imply either (a) that the three systems are autonomous and independent of one another or (b) that additional complexities not

Table 3.1 *Elements of a socio-ecological analysis of the Katrina disaster*

	Micro	*Meso*	*Macro*
Natural	Factors leading the storm to hit land near New Orleans.	Multi-year cycles affecting water temp in the Gulf of Mexico.	Climate change. Increased storm intensity.
Technological	Storm surge. Break in levee. Flooding of New Orleans.	Bureaucratic problems preventing proper design, construction and repair of the levee.	Economic forces affecting location of city. Ideology of environmental domination.
Social	Differential construction of victimhood. Breakdown of dual system of social order.	Social processes leading to spatial segregation.	Institutional racism. The heritage of slavery. Segregation.

captured by the table do not exist. The following subsections examine each of the three systems in turn.

Natural system: Hurricane formation, intensity and path

Like a leaf being steered by the currents in a stream, hurricanes are directed by the pressure ridges in existence at the time they form. Thus, at the micro level, that is during the few days involved with Katrina's existence, her path was determined by the surrounding flows in the troposphere. Once formed, a variety of physical factors affect a hurricane's intensity: moving over warm ocean water which provides energy for the hurricane; weak winds (because high winds bring up colder ocean water); low wind shear (because vertical shear can destroy the vortex); moist air in the hurricane's core; and avoiding land, which creates high frictional drag and deprives the storm of warm ocean 'fuel' (Emanuel, 2003). These factors came together during Katrina's existence. Specifically, after crossing Florida, Katrina encountered the Gulf Loop Current and Eddy Vortex, gained a huge amount of energy from the unusually warm waters, and shifted from a Category 3 to a Category 5 hurricane in just nine hours.

The micro-level physical forces affecting a hurricane's path and intensity, however, are embedded in larger meso-scale processes. Thus, for example, the Atlantic hurricane season, typically June through December, identifies the annual period when the conditions necessary for hurricane formation occur. The 2005 season was the most active season on record; a record 28 tropical and sub-tropical storms produced 15 hurricanes (4 of which reached Category 5 status). Hurricanes are also affected by longer scale rhythms. El Niño and La Niña refer to sustained changes in sea temperature in the South Pacific that last from two to seven years. The cold waters of La Niña affect atmospheric circulation patterns, increasing vertical wind shear in the Pacific and reducing wind shear in the Atlantic. Thus, the expanded area of low vertical wind shear during La Niña tends to produce more Atlantic hurricanes while fewer are produced during periods of El Niño. These ocean temperatures also influence where Atlantic hurricanes form, with more hurricanes forming in the deep Tropics from African easterly waves during La Niña. As a result, there is a much greater likelihood of them becoming major hurricanes (National Oceanic and Atmospheric Administration (NOAA), 2005).

At the even slower and larger macro scale, the Atlantic multidecadal oscillation affects sea temperatures and hurricane activity over the 40–80 year timeframe (Kerr, 2000). While the number of hurricanes is not changing, the last 30 years have witnessed a doubling of the proportion of very intense (Category 4 and 5 hurricanes) from roughly 17 per cent to 35 per cent of all hurricanes (Webster et al, 2005). Indeed, the change in hurricane intensity appears to be part of a process of energy redistribution associated with global climate change, driving heat from the tropics to the poles. In sum, a full understanding of the causal forces responsible for Katrina's intensity and path requires attention to a range of natural processes operating across a variety of spatial-temporal levels. The same is true of the technological system.

Technological system: The levees and other water control technologies

New Orleans is like a TV-dinner dish; much of the city is below sea level and protected from water by a set of levees including internal levees dividing the larger bowl into five separate compartments. Rain water aside, New Orleans is threatened by water from three distinct sources: the Mississippi River (effectively held in check by high quality levees constructed following the flooding of the city in 1927), Lake Pontchartrain, and storm surge from the Gulf of Mexico to the east and south. It is the lesser levees associated with the latter two that were breeched (van Heerden and Bryan, 2006, pp76–78).

At the micro level, Katrina's 10–20ft storm surge breached New Orleans' federally built levee system in 53 places. Eighty per cent of the city flooded with 87 per cent of the water originating from levee failures that totalled less than 400 yards along the drainage canals and approximately 650 yards on the Industrial Canal. Van Heerden and Bryan (2006, pp80–106) provide a detailed account of the micro processes associated with the storm surge and the sequence of levee breaches. In addition, the levees suffered from a variety of flaws – design, construction, maintenance, and so on – rooted in longer term (meso- and macro-level) processes.

The water control technologies in New Orleans developed over centuries in order to reclaim swamp land for urban expansion, to protect the city from flooding by the Mississippi and for hurricane protection. In the interest of saving space this section draws attention to a number of factors discussed further by Colten (2005), Kerstenbaum (2006), Woolley and Shabman (2007) and numerous others without analysing them in detail. Simply put, a number of bureaucratic problems adversely affected the design, construction and maintenance of the levee system. First, the original 1962 design for the Lake Pontchartrain and Vicinity Hurricane Protection Project was engineered to specifications for a 'standard project hurricane' rather than the less likely but stronger 'maximum possible hurricane' and in the following decades was modified a number times, often in response to public concern about particular suggestions, without an overall reassessment of the system's effectiveness. One of the design changes involved the use of an I-type floodwall that was cheaper to build and, purportedly, as strong as the earlier T-wall construction. Unfortunately, many of the I-type walls failed.

A second design consideration involved the Mississippi River Gulf Outlet (MR-GO). Construction of the MR-GO (1956–1965) overlapped the design phase of the hurricane protection system but little attention was paid to the impact of one project on the other. However, the MR-GO played a significant role in delivering Katrina's storm surge into New Orleans. Furthermore, a study by Hassan Mashriqui showing that the 'funnel effect' could increase the storm surge hitting the Intracoastal Waterway by 20–40 per cent, was ignored when the levees were constructed (van Heerden and Bryan, 2006, pp79–81).

Budgetary considerations were another factor. Since both the MR-GO and the Hurricane Protection Project were built by the Army Corps of Engineers, they competed for funds from a single budget. As of August 2005, fully 30 years after the original completion date, the percentage of completion of various hurricane protection project 'units' ranged from 65 to 98 per cent. These percentages had remained essentially constant since 1994. Moreover, much of the completed levee system was below design height specifications because the land had settled. Bureaucratic ineptitude and infighting also played a significant role in both the lack of completion and the insufficient maintenance of the levees. While the overall design was the responsibility of the Corps of Engineers, construction involved cost sharing arrangements among a welter of federal, state and local agencies, and the responsibilities for construction and maintenance were likewise decentralized (Horne, 2006).

In sum, despite warnings that the levee could fail, there were a large number of meso-level factors that, taken together, affected the viability of the levee system: (1) different organizations responsible for the design and the maintenance of the levee system (Cooper and Block, 2006); (2) under-funding of levee construction because of the dual responsibility (economic development, storm protection) of the builder and its relative emphasis on economic development (i.e. canals and dredging of waterways such as the MR-GO in order to facilitate ship transportation) (Horne, 2006); (3) cuts in federal funding from the Bush Administration due to federal concern with terrorist disasters rather than natural disasters (Cooper and Block, 2006), coupled with (4) the perception of New Orleans as an island of Democratic support in an otherwise Republican state (Horne, 2006). At the largest scale, it was the levees, pumps and drainage ditches that made the expansion of New Orleans (through the reclamation of land that was below sea level) possible and led to the creation of a city in a bowl.

However, these various water control technologies would not have been necessary if New Orleans had been located in a different place. Thus, at the macro level, it is necessary to understand how New Orleans came to be an inevitable city in an impossible location. The answer lies in the economic and ideological relation of humans and their environment. Water transport is the most economically efficient mode of transporting large volumes of product vast distances and, officially founded in 1718, the original city rested on a slip of high ground at the crossroads of three navigable bodies of water (Lake Pontchartrain, the Gulf of Mexico and the Mississippi River). As America settled the Ohio valley and Louisiana Territory in the 19th century, commerce along the Mississippi River increased and New Orleans grew appreciably, requiring repeated bouts of land reclamation and levee building. Thus, it was the human transformation of the natural environment that enabled the city to expand. What began with efforts to tame the Mississippi River (through flood control) and reclaim land (to enable the city to grow) gradually turned into a system also aimed at protecting the city from hurricanes. Additionally, as water was removed from the swamps to reclaim land, the dry land settled and the city continued to sink (McQuaid and Schleifstein, 2006, pp3–101). The wetlands, which serve as a natural buffer against storms, were further affected by various channels cut through them to facilitate oil development around the Gulf of Mexico (van Heerden and Bryan, 2006).

Underlying these developments was an ideology of environmental domination in which taming nature, rather than living in harmony with it, lay at the root of the city's physical vulnerability (Cutter, 2006). Thus, as with the natural system, we see that a full understanding of the details of the disaster requires one to attend not only to the micro-level processes directly responsible for the breech of the levees, but also to examine the meso- and macro-level process that also affected the viability of the storm protection system.

Social system

While adequate government response to Hurricane Katrina's impact in New Orleans may have minimized the ensuing social disruption, the chaos was an emergent phenomena embedded in existing processes and histories, not a failure of external social control. To understand the uniqueness of the social response, we must understand the uniqueness of the social context. We begin, as in the previous sections, at the micro level.

The key observation, documented in the first section, was the atypical nature of the social response to disaster evident in New Orleans. What accounts for this? Why did Katrina lead some citizens of New Orleans to exaggerate, rather than minimize, their social differences? Heeren's (1999) study of individuals affected by southern Californian fires provides some clues. He found that the identity of 'fire victim' became very salient, overriding other roles and statuses. This occurred in a series of stages. First, traditional identities were removed. Analogous to Goffman's (1961) concept of identity stripping, this involved direct experience with the disaster and the recognition of how profound its consequences would be. Second, the identity of 'fire victim' was constructed, initially as a self-definition and subsequently through processes involving reinforcement by the media and other institutions. Bonding occurred, as suggested by Durkheim (2001), because the disaster focused the emotions of a significant group (the fire victims) on a single object and its effects. The assumption was that, as victims of the same fire, they experienced the same emotions and, hence, saw themselves as sharing an identity-defining experience. Heeren also noted the importance of the insider/outsider boundary. This line of identity demarcation prevented non-victims from understanding what it meant to be victims. As a result, non-victims and victims treated each other differently. In short, Heeren shows that 'disaster victim' is a constructed identity and that the division between 'victim' and 'non-victim' is important.

But what, exactly, were the residents victims of? Viewed in one light, they were all victims of Katrina. There were a number of widely shared consequences of the hurricane: disrupted electricity, communication, transportation, and so on. On the other hand, there were two dramatically different disasters: one attributable to the gale force winds and another attributable to the flooding. The fact that the flooding was not uniformly distributed is significant in defining victims of the flood. Generally speaking, the areas of the city that flooded were poor and black, while the areas that did not flood were wealthy and white. The statistics are startling: 67 per cent of those affected by flooding were African-American, 28 per cent lived below the poverty line (of whom 84 per cent were black), 100,000 had no car, and therefore had no ability to flee the city when the storm hit (Strolovitch, et al, 2006). As a result, there was not a single uniform 'victim' identity. Instead, there were 'hurricane victims' and 'flood victims', and this division largely paralleled and reinforced existing social divisions rather than transcending them. Moreover, the initial response – in which the slow development of the flooding and the poor communication precluded a clear understanding of what was happening – served to heighten the sense of a people forgotten and to increase, rather than reduce, the sense of injustice. These perceptions were exacerbated by the government response. As noted by Alexander (2006): 'the initial response to the disaster ... harked back to an earlier time when civilian disasters were tackled in a paramilitary way under the assumption that the principal problem was how to restore law and order, rather than how to restore health, safety and dignity to the affected population.'

In addition, the mass movement of people, from the flooded areas to the Superdome, the Convention Centre and other high ground, disrupted the distinct structures of social order that characterized different parts of the city. While the Garden District and the French Quarter operated according to white middle-class assumptions (the police are your friend and you call them when there is trouble), other areas operated according to Anderson's code of the streets:

> *At the heart of the code is the issue of respect, loosely defined as being treated 'right' or being granted one's 'props' (or proper due) or the deference one deserves... (The code) is actually a cultural adaptation to a profound lack of faith in the police and the judicial system ... (who are) viewed as representing the dominant white society and not caring to protect inner city residents.* (Anderson, 1999, pp33–34)

In this world, respect is constantly being negotiated, slights are acted on and individuals who call the police are told to 'stop snitchin'.

Thus, at the micro-level, we see the operation of three distinct processes. First, there was the experience of the disaster. This was substantially affected by race and class factors. The 100,000 residents without their own reliable transport were largely left to find their own mode of evacuation. Similarly, the parts of the city that flooded were primarily black and poor. Thus, substantial race and class based differences existed in the immediate experience of the disaster. Second, there was the process of identity transformation that accompanied these experiences. While all residents of New Orleans came to see themselves in one way or another as victims of Katrina, the differences in their experiences were obvious and those divisions became the basis for drawing boundaries between insiders and outsiders (i.e. victim types). I suggest

(a) that this was not a tidy process, generating neatly distinct groupings of 'poor black flood victims' and 'rich white hurricane victims' but, rather, a very messy process which generated a variety of different victim identities, (b) that the nature of the victim identity constructed was rooted in personal experience and, hence, the differential impact of the disaster on various racial and class groups resulted, at the population level, in racial and class differences in the distribution of various victim identities and (c) a correspondence between the distributed impact of the disaster and previous historical events (with wealthy whites suffering less than poor blacks). Taken together, these created a situation rife with ambiguity – where any given pair of people may or may not see themselves as similar victims – and, hence, rife with the opportunity for miscommunication and conflict. Third, the differential impact of the flooding disrupted the normal spatial functioning of the city, with its division of social order, by forcing the poor, black residents of the flooded areas to abandon their area of the city in order to move to higher ground.

These three factors, the large-scale disruption of the spatial organization of the city coupled with communication ambiguities rooted in differential victim identities that, in some cases, connected and reinforced historical legacies of oppression, explain both the variety of social responses (from helping behaviours to asocial acts) and the specific form of some of the more puzzling asocial acts (like the Gretna Bridge incident and the 'big dumps'). As with the natural and technological systems, these micro-level processes do not provide the entire social picture. As Cutter (2006) notes, the division of the city into socially distinct regions was a product of past dynamics and their impact on the spatial and social evolution of the city: (a) migration from rural impoverished areas to the city followed by white flight to suburban communities, (b) construction of public housing in undesirable areas like reclaimed land, (c) loss of inner city employment opportunities, particularly low-skilled ones and (d) the concentration of the most impoverished in certain neighbourhoods, with little or no employment, poor education and little hope for the future. Particularly significant is Cutter's reference to reclaimed land, the low-lying land that flooded when the levees failed. Furthermore, at the macro-level, the spatial division has its roots in a legacy of historical oppression: slavery and segregation.

Discussion and Conclusion

To summarize, the unusual social strife that accompanied Hurricane Katrina was an emergent phenomenon rooted in the unexpected interaction of three distinct systems: natural, technological and social. If Katrina had not occurred or had taken a different path and hit land somewhere else, then the social strife in New Orleans would not have occurred. If the levees had not been breached and the city had not flooded, then the nature of the social response would have been very different. If New Orleans had a different social and political history, then the social strife would not have occurred or would have taken a different form. It is only by taking all three systems into account that one gets a full picture of what happened (and continues to happen) in New Orleans.

Moreover, a full explanation of these events requires attention to a spectrum of processes within each system ranging in scale from the short-term and micro to the long-term and macro. Despite the oversimplifications present in Table 3.1, the taxonomy draws attention to the range of phenomena necessary for understanding the unexpected nature of the post-Katrina New Orleans experience. Succinctly put, each of these processes was individually necessary. However, individually, these processes are not sufficient to understand the unexpected occurrence of civil strife. It is only through an account that incorporates the entire range of processes, across all three systems and all three spatial-temporal scales, that a sufficient explanation becomes visible.

In light of this account and the elements it emphasizes, it is useful to contrast Katrina with another disaster strongly associated with negative social consequences, the Buffalo Creek Flood. On February 26, 1972 a dam in the coal country of West Virginia collapsed and 132 million gallons of water and coal mining waste raced down the valley in the form of a 20–30 foot tidal wave, devastating 16 small communities, killing 125 and leaving 4000 homeless. Psychiatric examinations taken after the disaster found 93 per cent to be suffering an identifiable emotional disturbance. Erikson (1976) attributed the persistence of trauma and what he considers an unusual failure of time to heal the disaster's wounds to a lost sense of communality. Crudely put, Erikson characterized the local culture in terms of two competing character types – one independent and assertive and the other fearful and dependent – and argued that the loss of communality resulting from the flood's disruption of taken for granted community activated the latent potential of the weaker tendency and, hence, the local population experienced a tipping point and flipped into the despondent, dependent, fearful character mode.

Seen in the language of this chapter, Erikson both (a) treats the disaster as a system accident involving the interaction of technological and social systems rather than a component failure accident rooted completely in the failure of the dam and (b) recognizes the importance of social context (as evident in his reference to the specifics of Appalachian culture). Similarly, both accounts emphasize the scale of the disaster and the resulting impact on the conceptualization of victimhood. While Erikson does not produce a socio-ecological analysis of the Buffalo Creek disaster, it would be easy to do so as his account emphasizes a number of factors, both technological and social, operating on a variety of spatial-temporal scales.

That said, Erikson's account of social trauma in Buffalo Creek differs significantly from the Katrina case. First, Erikson's account treats the disaster as a cause of trauma; social divisions that did not exist prior to the disaster emerged in its wake. The Katrina account treats the social conflict that emerged in the wake of the disaster as a reflection of long-standing and persistent social divisions rather than their cause. Second, members of the Buffalo Creek community involved in the decisions creating and maintaining the dam were blamed for the tragedy and the mining company that employed them became the object of a lawsuit. While the flood control system in New Orleans was certainly riddled with flaws, these were (at least in part) the result of public input into the design.

Thus, culpability was neither as obvious nor as personal as in Buffalo Creek. As a result, where social divisions in the Buffalo Creek case involved allocation of blame, those in the New Orleans involved perceptions of differential exposure. Third, Erikson emphasizes the homogeneity of the residents and the existence of an holistic and all-encompassing sense of community prior to the flood. In contrast, the shared sense of community in New Orleans prior to Katrina was a thin veneer painted over fundamental divisions.

Finally, Erikson focuses on ongoing and persistent social divisions where the present account emphasizes micro-processes of social division immediately following the disaster. There is, however, an ongoing legacy from Katrina, both in the symbolic meaning different groups attribute to the disaster and the way it informs their lives and, as described by Barrios (this volume), in the ongoing disputes over how to rebuild New Orleans. Policies aimed at rebuilding the city need to take account not only of what has been learned about the dynamics of hurricanes (the natural system) and the failures in the technological system of hurricane protection, but also the historic social divisions that permeate the city and were exacerbated by the disaster (the social system).

References

Abbott, A. (2005) 'Process and temporality in sociology', in G. Steinmetz (ed) *The Politics of Method in the Human Sciences*, Duke University Press, Durham, NC, pp393–426

Alexander, D. (2006) 'Symbolic and practical interpretations of the Hurricane Katrina disaster in New Orleans', http://understandingkatrina.ssrc.org/Alexander/ accessed 8 July 2007

Ali, H. (2004) 'A socio-ecological autopsy of the e. coli O157:H7 outbreak in Walkerton, Ontario, Canada', *Social Science and Medicine*, vol 58, no 12, pp2601–2612

Anderson, E. (1999) *Code of the Streets: Decency, Violence, and the Moral Life of the Inner City*, Norton and Company, New York

Barton, A. (1969) *Communities in Disaster: A Sociological Analysis of Collective Stress Situations*, Anchor Books, Garden City, NJ

Brinkley, D. (2006) *The Great Deluge: Hurricane Katrina, New Orleans and the Mississippi Gulf Coast*, HarperCollins, New York

Colten, C. (2005) *An Unnatural Metropolis: Wresting New Orleans from Nature*, Louisiana State University Press, Baton Rouge, LA

Congleton, R. (2006) 'The story of Katrina: New Orleans and the political economy of catastrophe', *Public Choice*, vol 127, no 1, pp5–30

Cooper, C. and Block, R. (2006) *Disaster: Hurricane Katrina and the Failure of Homeland Security*, Times Books, New York

Cutter, S. 'The geography of social vulnerability: Race, class, and catastrophe', http://understandingkatrina.ssrc.org/Cutter/ accessed 8 July 2007

Douglas, M. (1966) *Purity and Danger: An Analysis of the Concepts of Pollution and Taboo*, Routledge, New York

Drabek, T. (1986) *Human System Response to Disaster: An Inventory of Sociological Findings*, Springer-Verlag, New York

Durkheim, E. (2001) *The Elementary Forms of the Religious Life*, Oxford University Press, Oxford

Emanuel, K. (2003) 'Tropical Cyclones', *Annual Review of Earth and Planetary Sciences*, vol 31, no 1, pp75–104

Erikson, K. (1976) *Everything in its Path: Destruction of Community in the Buffalo Creek Flood*, Simon and Schuster, New York

Fritz, C. (1961) 'Disaster', in Merton, R.K. and Nisbit, R.A. (eds) *Contemporary Social Problems*, Harcourt, New York, pp651–694

Goffman, E. (1961) *Asylums: Essays on the Social Situation of Mental Patients and Other Inmates*, Doubleday, New York

Heeren, J.W. (1999) 'Emotional simultaneity and the construction of victim unity', *Symbolic Interaction*, vol 22, no 2, pp163–179

Horne, J. (2006) *Breach of Faith: Hurricane Katrina and the Near Death of a Great American City*, Random House, New York

Kerr, R. (2000) 'A North Atlantic climate pacemaker for the centuries', *Science*, vol 288 no 5473, pp1984–1986

Kestenbaum, D. (2006) 'Why did the 17th Street Canal levee fail?', www.npr.org/templates/story/story.php?storyId=5418521 accessed 2 June 2008

Klinenberg, E. (1999) 'Denaturalizing disaster: A social autopsy of the 1995 Chicago heat wave', *Theory and Society*, vol 28, no 2, pp239–295

McQuaid, J. and Schleifstein, M. (2006) *Path of Destruction: The Devastation of New Orleans and the Coming Age of the Superstorms*, Little Brown, New York

National Oceanic and Atmospheric Administration. (2005) 'Frequently asked questions about El Niño and La Niña', www.cpc.ncep.noaa.gov/products/analysis_monitoring/ensostuff/ensofaq.shtml#HURRICANES accessed 5 July 2007

Perrow, C. (1999) *Normal Accidents: Living with High Risk Technologies,* Princeton University Press, New Haven, CT

Strolovitch, D., Warren, D. and Frymer, P. (2006) 'Katrina's political roots and divisions: Race, class, and federalism in American politics', understandingkatrina.ssrc.org/FrymerStrolovitchWarren/ accessed 8 July 2007

Solnit, R. (2009) *A Paradise Built in Hell: The Extraordinary Communities That Arise in Disaster*, Viking, New York

United States Congress (2006) 'A Failure of Initiative', www.gpoaccess.gov/katrinareport/fullreport.pdf accessed 15 July 2007

van Heerden, I. and Bryan, M. (2006) *The Storm: What Went Wrong and Why During Hurricane Katrina*, Viking, New York

Webster, P., Holland, G., Curry, J. and Chang, H. (2005) 'Changes in tropical cyclone number, duration, and intensity in a warming environment', *Science,* vol 309, no 5742, pp1844–1846

Woolley, D. and Shabman, L. (2007) 'Decision-making chronology for the Lake Pontchartrain and vicinity hurricane protection project', http://graphics8.nytimes.com/packages/pdf/national/20070711_HPDC.pdf accessed 15 July 2007

4
Conceptualizing Disasters as Extreme Versions of Everyday Life

Edward Woodhouse

Many chapters in this volume help locate the Katrina disaster within the broader context of disaster studies, both 'natural' and anthropogenic. This chapter contributes to that effort, but also aims more fundamentally to move the conversation out of what might be termed the intellectual ghetto of disaster/risk studies. Whether it is the 1998 train derailment in Eschede, Germany, the sovereign debt crisis roiling European financial circles in 2010, or the avoidable suffering associated with Hurricane Katrina or the Haitian earthquake, I argue that events known as 'disasters' are caused by many of the same precursor conditions that lead to less obvious outcomes in everyday life, outcomes which cumulatively may be even more destructive than the headline events.

Toward those ends, I pose and partially answer one deceptively simple question: What would it take to significantly reduce the probability and magnitude of especially bad technologically mediated outcomes, including weather-related disasters like Katrina? Trying to answer that question makes evident that the political, economic and social reshaping of the conditions leading to most disasters would entail the same types of reconstruction required for avoiding or significantly reducing many other harms to vulnerable populations.

I join Bowden (Chapter 3) in seeing the Katrina disaster as emerging not from specific component failures but from a systems accident (Perrow, 1984), from a set of natural, technological and social forces coupled in a complex system beyond anyone's complete comprehension. Neither the hurricane per se nor the failure of specific flood control components were enough to produce what came to be experienced as 'Katrina'; race, class and partisan politics were among other crucial contributors (Strolovitch et al, 2006). Whereas Bowden focuses on the social response to disaster, my analysis is confined to understanding how the forces leading to disastrous outcomes might be headed off or significantly reduced. To summarize the answer up front, albeit a bit too simply, I believe that natural and technological disasters together with many other unwanted social outcomes can be traced largely to failures to cope with uncertainty and with disagreement.

Towards symmetrical, integrated analysis of disasters

How might scholars proceed to think about hurricanes and other natural disasters in ways that:

- allow comparisons among many different kinds of disasters, including those normally termed 'technological' rather than 'natural';
- conceptualize disasters and other unwanted but less damaging outcomes along similar lines;
- analyse desired and undesired outcomes using the same variables; and
- frame the causal analysis so as to lead rather directly to proposals for mitigating the frequency and extent of future problems?

One way to approach the Katrina case in this symmetrical (Bloor, 1991) and integrated way is to think about what it takes to avert especially bad outcomes.

To put it tautologically, catastrophes are extreme, undesirable outcomes that have not been averted; conversely, when outcomes are more favourable something evidently has averted the potential unfavourable results. Thus, extreme weather events precipitate especially damaging outcomes only when government officials and others do not craft suitable building codes, do not plan for emergency transport and supplies, and so forth. What would it take to increase the odds of adequate preparation (and equivalent steps of other sorts), or, conversely, what goes wrong when severely unfavourable outcomes do occur?

One school of thought from political science holds that the key to reducing the frequency and magnitude of unfavourable outcomes is to arrange institutions and procedures so as to prevent big oversights and missteps from being taken in the first place, or at least from becoming entrenched. Under certain conditions, this can be achieved via incrementalist decision processes that take advantage of the inevitability and desirability of disagreement among diverse partisans (Dahl and Lindblom, 1953). One faction's big ideas often will be blocked or whittled down to size by other factions, yielding slower, smaller steps that produce smaller errors and more time for corrections based on learning from experience.

These ideas were refined in STS controversy studies starting with Collingridge (1980). Uncertainty has to be assumed in complex technological endeavours, he averred, but uncertainty per se is not the main problem; rather, the big opponent is inflexibility. Long lag times before errors can be discerned are one component of inflexibility – think Mississippi River-Gulf Outlet (MR-GO) canal. Inflexibility tends to arise when large organizations make centralized decisions without sufficient negotiation with a diversity of stakeholders (Collingridge, 1992). Civilian nuclear power ran afoul of this, as did system-built, high-rise public housing and many other technological endeavours gone awry.

The basic intent of the neo-incrementalist approach is to cope intelligently with the uncertainties and disagreements inevitable in most endeavours. Rather than shoving aside dissenting voices, actively incorporate a great diversity of perspectives into technological decision making. And rather than ignoring or denying uncertainties, assume that uncertainty is rife and errors are inevitable; build in redundancies, organizational slack and other strategies to cope with undesirable events as they become evident and to actively learn from experience.

Whether the problem is earth/wind/water/temperature-mediated natural disasters or engineering failures such as the Hyatt Regency catwalk collapse (Martin, 1999); whether the scale is large and acute as in Bhopal (Fortun, 2001) or small and chronic as in the shortage of facilities for skateboarders (Harnick and Gentles, 2009); and whether the outcome is terrible or wonderful, techno-social outcomes can be analysed in terms of how well decision making coped with uncertainties and with disagreements. The quality of technological and other outcomes correlates fairly well with the degree of adherence to a set of decision-making practices I refer to as intelligent trial and error (ITE) – practices that reduce the frequency and severity of major errors, and thereby often evolve outcomes considered relatively successful (Woodhouse and Collingridge, 1993). Consider with me how selected facets of ITE apply to the Katrina disaster, and how this conceptual approach allows Katrina and other disasters to be understood as extreme manifestations of phenomena occurring ubiquitously (and failing to occur) in everyday life.

Timely and effective deliberation?

Historians and sociologists find substantial malleability early in the shaping of technological trajectories, but stabilization and even closure typically occur at some point (Bijker et al, 1987). Appropriate design usually requires watching young technologies and inflecting their trajectories – a form of learning by doing – because in the early stages there is rarely sufficient knowledge or an attentive constituency of would-be adversaries. Yet by the time that the requisite attention becomes available, technological momentum may have set in – the Collingridge dilemma (1980). Synthetic biology and robotics are two contemporary examples where momentum is building prior to effective deliberation.

Some technological designs can be reopened for reshaping, as occurred with development of the mountain bike several generations after ‘the bicycle’ originally stabilized. New materials became available, new opportunities arose, and other changes made it feasible to redo bike frames, suspensions, gearings and tyres in non-trivial ways (Buenstorf, 2003). The results improved outcomes for millions of riders because their ‘representatives’ (mostly avid cyclists-turned-small-business-proprietors) learned from experience and renegotiated bicycle design.

For nuclear power, in contrast, to date there has been a moratorium on new reactors in many nations but no significant effort since the Three Mile

Island and Chernobyl accidents to address public objections through fundamental redesign. This has been true despite price and supply issues around petroleum coupled with enhanced recognition of the health and environmental costs of coal (Meadowcroft and Langhelle, 2010). One need not make partisan claims on behalf of civilian nuclear energy to note the lack of public deliberation about the possibility of a second nuclear era based on 'inherently safe', small reactors that cannot suffer meltdowns (Morone and Woodhouse, 1989). Lidsky and others began developing such designs a quarter of a century ago, and Babcock and Wilcox engineers have laboured in obscurity prototyping small reactors ever since.

With the Obama administration's newfound openness to nuclear power as part of the effort to reduce carbon dioxide emissions, a number of US electric utility companies are now joining in a licensing application to the US Nuclear Regulatory Administration (Smith, 2010). This is likely to provoke renewed controversy, and may lead to close scrutiny of the new designs by environmental groups and other advocates and adversaries. This reopened deliberation, if it happens, will be coming at least two decades later than would have been feasible if nuclear scientists, social scientists, journalists and environmental groups had known of the technical potentials and pushed for timely debate.

In the case of Katrina, informed and widespread debate no doubt ought to have occurred prior to construction of the now-closed MR-GO shipping canal that brought the storm surge into New Orleans. However, the canal was proposed a half century ago when the field of ecology, environmental NGOs and public interest law were barely in their infancy; so there was radically less capacity to mount a genuine inquiry. An absence of contested deliberation likewise accompanied the diffusion of synthetic organic chemicals in the 1920s–1950s, the Great Bandwagon Market for light water reactors in the 1960s, and the Nixon administration's 1971 decision to give nuclear missiles multiple independently targetable warheads (which made arms control much more difficult). Indeed, technological innovation not atypically has proceeded in ways that Langdon Winner (1977) terms 'technological somnambulism' – a lack of informed, thoughtful, deliberate choice.

MR-GO in that sense was in the mainstream, not something unusual. Even with the development of some of the enabling conditions for timely and knowledgeable contestation, robotics and other epochal technologies of the contemporary era nevertheless are proceeding without much explicit pre-market deliberation. Social and ethical issues around nanotechnology, for example, are being *studied* under large government grants; but neither the US nor the EU has a regulatory scheme in place despite the rapid proliferation of nanotechnoscientific capacities and products (van Calster, 2006). And press coverage tends to highlight gee whiz features and economic competitiveness, rather than aiding sober assessment. Again, this is not out of step with everyday life; social psychologists could testify at considerable length to the fact that both expert and lay thinking tends to be plagued by a combination of cognitive, emotional and situational biases and impediments (Tetlock, 2005).

Some deliberative shortcomings stem from the fact that it often takes a lot of expertise, funding, organization and access to participate meaningfully – and those requisites generally are in short supply. The Mackenzie Valley Pipeline hearings regarding the Trans-Alaska Pipeline worked as well as they did largely because the administrative law judge took it upon himself to bring the hearings to many different stakeholders in Canada (Berger, 1977). Following the Danish lead, the UK, US and other political systems are experimenting with consensus conferences, deliberative polling and other deliberative forums intended to increase citizen knowledge and participation in technological decision making (Fishkin, 1991; Hamlett et al, 2008). No nation's government officials yet have begun to share real authority with such new forms of semi-representative deliberation.

In both private and public decision making, from the very small scale to the very large, deliberation often is rather narrow. The MR-GO canal advocates focused more on shipping than on flood control, and more on engineered levees than on natural barriers to wind and wave (Hoekstra and Hung, 2003). In designing the space shuttle, National Aeronautics and Space Administration (NASA) officials accommodated the Department of Defence in ways that led to long-term shortcomings for civilian missions (Kay, 1994). Deliberations over fire safety at the World Trade Center were narrowed by the fact that it was a Port Authority Project and under the arcane laws of New York State the Authority was not subject to city building regulations (Glanz and Lipton, 2003). Hence, the building was constructed in ways that violated what turned out to be crucial parts of the local building code that some observers believe could have allowed the towers to remain standing long enough to evacuate most of those who perished. Breadth and intensity of deliberation/oversight frequently is sacrificed on the altar of doability and quickness.

Local knowledge is often screened out of deliberations as a by-product of the marginalizing of some affected interests. Politics is about 'mobilization of bias', as Schattschneider (1960) put it, and those doing the mobilizing frequently dispossess others of the opportunity to bring their considerations to bear. This often reduces the intelligence of a decision, because proponents can duck challenge by critics.

The reduced intelligence is far from random. Whereas representative democracy sometimes is assumed to provide more or less automatically for fair sharing of influence, major civilian technologies actually emerge primarily from the research and business sectors, where representative democracy normally intrudes indirectly and feebly. Government officials play the roles of cheerleader, funding agent or bystander more often than 'decision maker'. This is due partly to the privileged position of business, whereby the relevant industry executives generally not only enjoy discretionary authority over what to bring to market but also have disproportionate funding, organization, access and other resources to deploy in influencing government officials where necessary (Lindblom, 1977). Even buyers in market exchange normally have little incentive to take account of their effects on third parties, and so-called private-sector decision making therefore rarely gives appropriate weight to public considerations (Lindblom, 2000). Moreover, the privileged position of business gives corporate elites and their government/society allies a unique position from which to communicate about what business 'needs' to keep the economy out of recession. Support for construction in coastal zones is a tangible example.

US politics also is skewed by gender, ethnicity and socio-economic class more than that of most other affluent nations, and Louisiana of course exemplifies these tendencies to an unusual degree. The extent of inequality is markedly lower in some political jurisdictions (e.g. The Netherlands – Milanovic and Yitzhaki, 2002) than in others. But in no nation is there an influence pattern around technological shapings that looks very much like what a random sample of humanity would comprise (Carson and Martin, 1999). Congressional hearings tilt heavily toward testimony from researchers and business executives, and regulatory agencies using advisory boards do not come close to genuine representativeness, despite various legal mandates (Jasanoff, 1990). Think of nanotechnoscience in our era: The National Nanotechnology Initiative in the US was originally suggested and approved by a narrow circle in the Office of Science and Technology Policy and the White House, and subsequently sanctioned by congressional 'hooray' (Jenkins, 2002). Other interests such as public health were marginalized, and had to be shoehorned in later once the health risks of nanoparticles became apparent. Shaping of the Katrina outcomes was not much different from this sort of technological business-as-usual.

Even technological decisions not exactly made in secret usually are not highly transparent. Transparency allows outsiders to monitor decision making, partially compensating for shortcomings in diversity of participation. When insiders know they are being watched, they are less likely to entirely ignore what they perceive as the watchers' concerns. That no one knew whether the pilings anchoring the levee walls were 16 feet or 24 feet or some other depth is one indicator of lack of transparency. Again, this is not atypical, even for subjects on which there are public hearings and nominal transparency, such as contemporary nanotechnology, for a good case can be made that hearings serve a largely dramaturgical and symbolic function (Edelman, 1971; Hilgartner, 2000). Committee markups where real decisions sometimes are made are not open to journalists or the public, and many events occur far from the public eye. Outside of public notice and probably little attended even by the board of education was the location and design of school bus garages in New Orleans, which belatedly became a major issue when many dozens of buses had to be replaced at a cost of $100,000 each following the flood.

In addition to transparency and fair participation, or their opposite, decision processes allocate the authority to choose which events warrant specific actions. One facet of such choice regarding Katrina concerned post-war New Orleans housing that was constructed for the most part without the time-tested precautions used in older 'shotgun'-style housing. Whereas the traditional methods raised the main part of the house well off the ground, built with cypress wood that does not rot, and spaced the floor boards widely to facilitate floodwaters draining out, the newer housing used the same slab-on-grade methods found in tract homes built for very different conditions. Although the particulars are unique to coastal Louisiana, the underlying decision-making process is widespread: delegate to 'private' actors and institutions the responsibilities for making important decisions with fateful public consequences. Unless incentives are in place to induce private decision makers to incorporate appropriate public considerations, this delegation normally will result in misshapen technologies.

Prudent coping with uncertainties

Prudent coping with uncertainties is sometimes still possible even when stakeholders are marshalled out of decision making, and deliberations are untimely and/or too narrow in scope. Collingridge described the basic dilemma confronting innovation this way:

> *Attempting to control a technology is difficult, and not rarely impossible, because during its early stages, when it can be controlled, not enough can be known about its harmful social consequences to warrant controlling its development; but by the time these consequences are apparent, control has become costly and slow.* (1980, p19)

This predicament has become somewhat familiar via experience with successive waves of emerging technologies, and scholars have uncovered a set of strategies useful for coping with uncertainties (Woodhouse and Collingridge, 1993). Although avoiding mistakes is too much to expect, it is possible to proceed via 'intelligent trial and error' rather than by blind trial and error (Lindblom and Woodhouse, 1993). When those shaping technological trajectories look upon 'engineering as social experimentation', they stand a reasonable chance of reducing the magnitude of errors while speeding up learning from feedback (Martin and Schinzinger, 1996).

Precautionary design before actual operation is one ingredient. In the early years of recombinant DNA experiments, microbiologists meeting at Asilomar decided to require containment laboratories for risky experiments (Krimsky, 1984). There were reasonable people at that time who considered the special measures an unnecessary expense, and subsequent experience by and large proved the sceptics to have been correct. Nevertheless, many people preferred to err on the side of caution, in effect purchasing insurance that never had to be collected on. Initial precautions make especially good sense when products or planning strategies are likely to become entrenched once released to market. Similarly, the (lack of) hurricane preparedness in New Orleans became more entrenched and thus more difficult to correct as the city grew.

Another example of initial precautions is that nuclear plants in the US are required to have evacuation plans in case radiation releases threaten local communities. The Shoreham Nuclear Plant on Long Island was constructed at a cost of some six billion dollars in 2008, but was mothballed before ever operating because New York State officials did not find the evacuation plans believable. Given the traffic patterns, there was no credible way to evacuate Long Island in a timely way. New Orleans of course faces a similar constraint in terms of geography and transport corridors, yet the area has effectively been declared as operational despite the threat from rising waters that in some respects mirrors the threat from Shoreham.

A related facet of prudent planning involves erring on the side of caution in estimating threats and coping resources. For all their problems, nuclear reactors were well designed in this regard in that there were back-up sources of electricity in case one set of power lines failed. And there was a diesel generator as a further back-up, and a second generator in case the first one failed. 'Defence in depth' is the term nuclear engineers use to describe such multiple back-ups. What if New Orleans' officials had applied that tactic in estimating the buses and other transport requirements, or in their estimations of necessary supplies of clean drinking water, sanitation facilities, emergency housing and emergency food? It is so far from everyday practice – and not just in New Orleans but in disaster preparedness in most locales – that it is difficult even to imagine duplicating the careful approach of the nuclear engineers.

However, when highly unfavourable outcomes are averted, as occurred with the successful response to a potential disaster during the 1988 flash flood in Nimes, France, one almost always finds reasonably good adherence to at least some of the decision strategies conducive to coping with uncertainties and disagreements (Maréchal et al, 2008). The Nimes flooding was radically smaller than Katrina, so a direct substantive comparison would be inapt; but the technocratic approach of French authorities meant that there was a capacity to provide aid quickly, because they had erred on the side of caution in arranging a nation-wide emergency plan. The German capacity for moving reverse osmosis water plants to disaster stricken countries likewise is a case of stockpiling more supplies than may actually be needed. These successful European examples did not emerge from boutique 'participation' or planning exercises; instead, the everyday operation of government and culture effectively represented a wide set of stakeholders who potentially could need aid in a crisis, and the negotiations establishing the programmes were both highly competent and highly motivated.

Preparation for learning from experience

Given the inevitability of unexpected consequences, learning from experience is the main way that socio-technical activities become successfully shaped and reshaped over time. Yet learning is not as easy as the phrase 'trial-and-error learning' makes it seem. 'Following September 11, 2001, public officials everywhere promised that the nation would learn the painful lessons the terrorist attack taught. But Hurricane Katrina not only revealed that we have failed to learn, it also showed that we have yet to build the capacity to deal with costly, wicked problems that leave little time to react' (Kettl, 2006, p273). As explained by Frickel and Vincent (Chapter 1), disasters and responses to disasters are partly rooted in organizational structurings. Both Homeland Security, in general, and the Federal Emergency Management Agency (FEMA), in particular, have organizational cultures and procedures that make it difficult to accomplish non-routine tasks, or even routine ones. But this is not atypical of large organizations. As human factors engineer Richard Hornick found while consulting for the Nuclear Regulatory Commission, complacency about safety was rife in the organization and actually became built into key pieces of equipment. For example, he found 'the word *pump* designated four different ways on the same control panel – PUMP, PP, PMP and PU – even where there was adequate space available so as not to require abbreviation at all. Another panel contained the colour red to signify three different states – emergency, warning/marginal, and normal/go!' (Hornick, 1987, p5).

The Presidential Commission on the Three Mile Island nuclear mishap found that those with responsibility for plant safety lacked relevant knowledge, utility representatives made inaccurate statements about the accident and about the plant more generally, and there was non-trivial discord within the Nuclear Regulatory Commission. In the preface to its report, the Commission acknowledged that for four days following the accident, 'the extent and gravity of the accident was unclear to the managers of the plant, to federal and state officials, and to the general public' (Kemeny et al, 1979, pxi) – a time period roughly equivalent to the initial confusion in New Orleans.

Another presidential commission, this one concerning the space shuttle *Challenger* disaster, found more or less the same phenomena at NASA – what they called a 'silent safety program'. In an agency employing tens of thousands including those working for industrial firms under contract with NASA, only about 20 people were assigned to work for the chief engineer at NASA headquarters who had overall responsibility for safety, reliability and quality assurance. And the commission members were astonished to learn that no one in that office actually spent more than quarter of their time working on safety issues. There also were safety programmes at the Johnson, Marshall and Kennedy Space Flight Centers, of course, but the commission found that safety officers lacked authority and were located at points in the organizational structure where they had neither operating authority or even independence. Illustrating the point was the fact that the fateful teleconference calls between Marshall and Thiokol staff prior to the Challenger launch decision did not include any NASA engineer whose primary responsibility was assuring safety.

FEMA, from its outset, had to report to 20 different congressional committees, which led to:

> *'Pulling ... in competing directions... (one of many) examples of executive reorganization that ... fails to gain traction because congressional oversight and appropriation responsibility is left unchanged. Executive reorganization (also) disrupts relationships with state and local government during the implementation phase'* (Balogh et al, 2002, p7).

However competent the staff and legislators of any given committee, the overall congressional structure is off limits for serious scrutiny and authoritative intervention by outsiders. So, the odds are not good of the US Congress being able to reflexively learn from its collective mistakes. Even replacing every legislator would not necessarily change the organizational culture and structure, and certainly would not change the multiple veto points at which governance can be slowed or diluted in the US tripartite system.

It is obvious on reflection that failure to learn is not unusual (Edmonson, 1996), and facing up to the fact ought to cast disaster studies and preparations in a somewhat different light. To the extent that Katrina repeats patterns from earlier tragedies, it means that something was not learned from earlier post-disaster analyses. This should alert us to the danger (and probability) that our own ways of thinking may not prove any more helpful to generalized, long-term learning than did earlier space shuttle (Columbia Accident Investigation Board, 2003), Bhopal (Fortun, 2001), World Trade Center (Glanz and Lipton, 2003) or Hurricane Camille (Zebrowski and Howard, 2005) inquiries.

A shortcoming in many post-disaster analyses such as The White House's (2006) report on Katrina is a tendency to assume that responsible parties will learn pretty much automatically from eminently sensible recommendations that the authors are about to convey. Assuming that residents of coastal Louisiana are no less intelligent than the rest of us, substantial barriers must exist to translating their latent need for safety into manifest deliberation and action. This suggests decision makers at all levels should not assume that lessons people learn will somehow make sensibility more common in the future.

Learning from experience obviously requires a way of obtaining relevant, timely, multifaceted information, and the odds of this increase as monitoring becomes more extensive, better-funded, and involving more diversity in the monitors' backgrounds and interests. As the falsified scientific reports from the Yucca Mountain nuclear waste repository illustrate, no set of insiders can be fully trusted (Macfarlane and Ewing, 2006). There simply is no substitute for multiple partisans with different biases observing and interpreting early trials of emerging technologies. In the case of leukaemia at Woburn, Massachusetts, lay people were indispensable in uncovering and publicizing the problem – a kind of 'popular epidemiology' (Brown, 1987), but their efforts came too late to prevent the tragedy. With the decline in civic culture in many political systems, such diversified monitoring seems less and less realistic as a standard practice.

Virtually no major technological activity anywhere in the world has invested appropriately in monitoring, so the failure to monitor Katrina dikes and pumping capacities is far from aberrant. But monitoring is improving here and there, as in required annual reviews of the US nanotechnology programme (Bucher et al, 2003). However, the evaluation is supposed to be done by the National Nanotechnology Initiative (NNI) steering committee itself, which obviously involves a conflict of interest. Self-evaluations tend to result in self-fulfilling assessments, usually from a narrow perspective. Rigorous external monitoring, performed by an independent panel, working with ample funds and staff would be in the spirit of the checks and balance that are a celebrated feature of the US Constitution. It might be considered utopian to expect such monitoring, but without it there is no realistic prospect of learning from experience either in large matters such as New Orleans' vulnerability or the European sovereign debt crisis, or in smaller matters such as the Toyota automobile recalls that were delayed in part by the culture of secrecy within many Japanese corporations. Well funded, expert, multipartisan monitoring is rare, and hence exemplary outcomes are rare.

Incumbents tend to compound inadequate monitoring with a tendency to stick with existing ways of doing things, resisting appeals to correct what others perceive as errors. To insiders, error correction means spending extra money and interfering with established routines and production schedules. Imagine the school board receiving an expensive proposal to move the bus garages at a time when teachers were being laid off for lack of funding! To improve the chances of policy learning, however, advanced democratic systems would build into policy trials incentives for those with authority to revise early trials as experience accumulates. Merely raising this issue on scholarly and activist agendas might constitute something of a breakthrough – and thus exemplifies the core spirit of this chapter in the sense that Katrina is more similar than different from other technological problematiques.

It also deserves noting that poor outcomes sometimes are deliberate, though it usually is difficult to know what fraction may fall in this category. Neglect of public schools, water systems and other government services can bolster the case for privatization, and both pro-business officials and business executives donating to campaigns know this. Post-Katrina New Orleans is one of the largest experiments with privatized schools in the US. Prior to Katrina many public schools had no working bathrooms and the kids used port-o-potties, whereas some of those same schools now feature soccer teams and fancy uniforms with a notably whiter student body.

Resistance to error correction emanates partly from expense. Even starting with a clean slate as in nanotechnology, where policy creativity is easier to imagine than in redress following a Katrina, no effective 'insurance' strategies typically are adopted. An exception is the excise tax on sales of gasoline, funding a Highway Trust Fund that makes it easier to pay for maintenance on roads. More commonly, reluctance to prepare financially for problems shows up about equally in emerging technologies, in disaster preparedness, and in everyday decisions such as whether to purchase long-term care insurance.

Conclusion

In closing, I have suggested that the same analytical lenses be used to understand disasters and other less dramatic social outcomes. More specifically: (1) one ought to study wind/water/earth natural disasters and technological disasters using similar lenses; (2) disaster analyses ought to use a framework that can also be applied to everyday technological outcomes; (3) desired outcomes ought to be explained symmetrically in the same terms as undesired outcomes; and (4) the causal analysis of problems ought to lead directly to proposals for improvements.

With respect to the Hurricane Katrina disaster, I have argued that it emerged from the same basic decision-making processes and strategies that govern other facets of techno-social life in the US: lack of the early and ongoing, multifaceted inquiry, contestation and renegotiation that gives flawed humans in flawed organizations a reasonable chance of catching and correcting serious errors of omission and commission. Because the Katrina catastrophe was concentrated and dramatic, whereas ordinary mini-catastrophes occur in small and widely distributed increments, the similarities are easy to miss. But the same processes and institutions work to create vulnerable populations, design the built environment carelessly with respect to public values, and make it easy for corporations and markets while placing barriers in the way of preventive governmental action. Inquirers need to face up to the fact that there is nothing all that special about the production of disasters, because lots of small, everyday problems add up to very significant problems for a great number of people and communities, and pretty much the same political and economic forces construct both the smaller problems and the larger ones.

References

Balogh, B., Grisinger, J. and Zelikow, P. (2002) *Making Democracy Work: A Brief History of Twentieth Century Federal Executive Reorganization*, Miller Center of Public Affairs, University of Virginia, millercenter.virginia.edu/pubs/homeland_security/full_report.pdf, accessed 30 July 2006

Berger, T. (1977) *Northern Frontier, Northern Homeland: Mackenzie Valley Pipeline Inquiry Report*, 2 vols, Ministry of Public Works and Government Services, Ottawa, Canada

Bijker, W., Hughes, T. and Pinch, T. (1987) *The Social Construction of Technological Systems*, MIT Press, Cambridge, MA

Bloor, D. (1991) *Knowledge and Social Imagery*, 2nd edn, University of Chicago Press, Chicago, IL

Brown, P. (1987) 'Popular epidemiology: Community response to toxic waste-induced disease in Woburn, Massachusetts', *Science, Technology, & Human Values*, vol 12, nos 3–4, pp78–85

Bucher, P., Birkenmeier, B., Brodbeck, H. and Escher, J. (2003) 'Management principles for evaluating and introducing disruptive technologies: The case of nanotechnology in Switzerland', *R&D Management*, vol 33, pp149–163

Buenstorf, G. (2003) 'Designing Clunkers: Demand-side Innovation and the Early History of the Mountain Bike', in M. Stanley and U. Cantner (eds) *Change, Transformation, and Development*, Springer, Dordrecht, pp53–70

Carson, R. and Martin, B. (1999) *Random Selection in Politics*, Praeger, New York

Collingridge, D. (1980) *The Social Control of Technology*, St. Martin's Press, New York

Collingridge, D. (1992) *The Management of Scale: Big Organizations, Big Decisions, Big Mistakes*, Routledge, London

Columbia Accident Investigation Board (2003) *Columbia Accident Investigation Board Report*, 6 vols, US National Aeronautics and Space Administration, caib.nasa.gov/, accessed 1 November 2006.

Dahl, R. and Lindblom, C. E. (1953) *Politics, Economics, and Welfare*, Harper & Brothers, New York

Edelman, M. (1971) *Politics as Symbolic Action: Mass Arousal and Quiescence*, Markham Press, Chicago, IL

Edmondson, A. (1996) 'Learning from mistakes is easier said than done', *Journal of Applied Behavioral Science*, vol 32, pp5–28

Fishkin, J. (1991) *Democracy and Deliberation: New Directions for Democratic Reform*, Yale University Press, New Haven, CT

Fortun, K. (2001) *Advocacy after Bhopal: Environmentalism, Disaster, New Global Orders*, University of Chicago Press, Chicago, IL

Glanz, J. and Lipton, E. (2003) *City In the Sky: The Rise and Fall of the World Trade Centre*, Times Books, New York

Hamlett, P., Cobb, M. and Guston, D. (2008) *National Citizens Technology Forum: Nanotechnologies and Human Enhancement*, The Center for Nanotechnology in Society, Report #R08-0002, Arizona State University, Phoenix, AZ

Harnick, P. and Gentles, C. (2009) 'Coming to a city near you: Skate parks', *Parks & Recreation*, vol 44 (May), pp34–38

Hilgartner, S. (2000) *Science on Stage: Expert Advice as Public Drama*, Stanford University Press, Stanford, CA

Hoekstra, A.Y. and Hung, P.Q. (2003) 'Virtual water trade: A quantification of virtual water flows between nations in relation to international crop trade', Virtual water trade proceedings of the international expert meeting on virtual water trade, *Value of Water Research Report Series*, no 12, pp25–48

Hornick, R. (1987) 'Dreams, design, and destiny', *Human Factors*, vol 29, pp111–121

Jasanoff, S. (1990) *The Fifth Branch: Science Advisers as Policymakers*, Harvard University Press, Cambridge, MA

Jenkins, D. (2002) 'Broken in Midstride: Space Shuttle as Launch Vehicle', in R. Launius and D. Jenkins (eds) *To Reach the High Frontier: A History of U.S. Launch Vehicles*, University Press of Kentucky, Lexington, KY

Kay, W. (1994) 'Democracy and super technologies: The politics of the space shuttle and space station freedom', *Science, Technology & Human Values*, vol 19, pp131–151

Kemeny, J., Babbitt, B., Haggerty, P., Lewis, C., Marks, P., Marrett, C., McBride, L., McPherson, H., Peterson, R., Pigford, T., Taylor, T. and Trunk, A. (1979) *The President's Commission on the Accident at Three Mile Island*, US Government Printing Office, Washington DC

Kettl, D. (2006) 'Is the worst yet to come?', *Annals of the American Academy of Political and Social Science*, vol 604, pp273–287

Krimsky, S. (1984) *Genetic Alchemy: The Social History of the Recombinant DNA Controversy*, MIT Press, Cambridge, MA

Lindblom, C. E. (1977) *Politics and Markets*, Basic Books, New York

Lindblom, C. E. (2000) *The Market System: What It Is, How It Works, and What to Make of It*, Yale University Press, New Haven, CT

Lindblom, C. E., and Woodhouse, E. J. (1993) *The Policy-making Process*, 3rd edn, Prentice Hall, Englewood Cliffs, NJ

Macfarlane, A. and Ewing, R. (2006) *Uncertainty Underground: Yucca Mountain and the Nation's High-Level Nuclear Waste*, MIT Press, Cambridge, MA

Maréchal, J., Ladouche, B. and Dörflinger, N. (2008) 'Karst flash flood in a Mediterranean karst: The case of Fontaine de Nîmes', *Engineering Geology*, vol 99, no 3, pp138–146

Martin, M. and Schinzinger, R. (1996) *Ethics in Engineering*, 3rd edn, McGraw Hill, New York

Martin, R. (1999) 'Hyatt Regency Walkway Collapse, Kansas City, Missouri, July 17, 1981', www.eng.uab.edu/cee/faculty/ndelatte/case_studies_project/Hyatt%20 Regency/hyatt.htm#Causes, accessed 27 February 2009

Meadowcroft, J. and Langhelle, O. (2010) *Caching the Carbon: The Politics and Policy of Carbon Capture and Storage*, Edward Elgar Publishers, Cheltenham

Milanovic, B. and Yitzhaki, S. (2002) 'Decomposing world income distribution: Does the world have a middle class?', *Review of Income and Wealth*, vol 48, pp155–178

Morone, J. and Woodhouse, E. J. (1989) *The Demise of Nuclear Energy: Lessons for Intelligent, Democratic Control of Technology*, Yale University Press, New Haven, CT

Perrow, C. (1984) *Normal Accidents: Living with High-Risk Technologies*, Basic Books, New York

Schattschneider, E. (1960) *The Semi-Sovereign People: A Realist's View of Democracy in America*, Holt, Rinehart and Winston, New York

Strolovitch, D., Warren, D. and Frymer, P. (2006) 'Katrina's political roots and divisions: Race, class, and federalism in American politics', *Understanding Katrina: Perspectives from the Social Sciences*, 11 June, Social Science Research Council, understandingkatrina.ssrc.org/FrymerStrolovitchWarren/, accessed 8 July 2007

Smith, R. (2010) 'Small reactors generate big hopes', *The Wall Street Journal*, 18 February, ppA1, A16

Tetlock, P. (2005) *Expert Political Judgment: How Good Is It? How Can We Know?*, Princeton University Press, Princeton, NJ

van Calster, G. (2006) 'Regulating nanotechnology in the European Union', *European Environmental Law Review*, August/September, pp238–247

The White House (2006) *The Federal Response to Hurricane Katrina: Lessons Learned*, Report submitted by the Assistant to the President for Homeland Security and Counterterrorism, February, Washington DC, www.whitehouse.gov/reports/katrina-lessons-learned/, accessed 9 November 2006

Winner, L. (1977) *Autonomous Technology: Technics-out-of-control as a Theme in Political Thought*, MIT Press, Cambridge, MA

Woodhouse, E. J. and Collingridge, D. (1993) 'Incrementalism, Intelligent Trial-and-Error, and the Future of Political Decision Theory', in H. Redner (ed) *An Heretical Heir of the Enlightenment: Politics, Policy and Science in the Work of Charles E. Lindblom*, Westview Press, Boulder, CO

Zebrowski, E. and Howard, J. (2005) *Category 5: The Story of Camille – Lessons Unlearned from America's Most Violent Hurricane*, University of Michigan Press, Ann Arbor, MI

Part II
Relocation, Rebuilding and Recovery Concerns

5
Mind Maps, Memory and Relocation after Hurricane Katrina

Kathryn Henderson

> *Among the 2000 Katrina evacuees housed in our university gymnasium, we try to find a house for a family of four grandmothers. They reject offers from agencies to place them in a retirement centre because they refuse to be separated from one another and a grandson with them. They like a house we find but ask, 'Where is the grocery store? drug store? bus stop?' They reject the house because none of these are close telling us their local Walgreen's clerk knew them by name and the details of their medical needs.*
>
> *After the evacuation of New Orleans our research team seeks feedback on a computer mapping prototype from social workers in Houston who are themselves Katrina survivors. The prototype links photos of New Orleans homes via GPS to a Google map of the city, providing a space where evacuees can provide information of their choice so people can search by neighbourhood for former residents. When we project the prototype the first thing someone says is, 'Where is the Seventh Ward?' The Google map is unfamiliar to the city's former residents. It is not the way they know or navigate their home space.*

Two seemingly mundane themes: (1) the importance of recreating in relocation the proximity of everyday needs sources in a manner echoing that of pre-disaster neighbourhoods; and (2) the situational specificity of hurricane survivors' local spatial and navigational knowledge of their evacuated neighbourhoods came up repeatedly in conversations with Katrina survivors in our mid-Texas college town and in New Orleans where they had returned from Houston and Dallas. Actions rejecting unfamiliar mappings and proposed residences needing independent transport for household resources, show hurricane survivors' attempts to recreate quasi-familiar navigable social space in a cognitive and physical landscape resembling their knowledge structure of pre-Katrina neighbourhoods. These were neighbourhoods of extended kinship and fictive-kinship networks which included centres for socializing and exchanging information, and commercial chain clerks who treated a person like a neighbour. Such situated, local knowledge is its own way of knowing. It is tacit, site-specific, and not captured in the Cartesian planes of commercial maps and GPS (global positioning system) coordinates.

Based on interviews and mapping exercises with Hurricane Katrina survivors, this study employs hand-drawn maps of neighbourhoods of origin in comparison with sparse sketches of relocation neighbourhoods and commercially produced maps of New Orleans to examine the situatedness of local spatial knowledge in daily interaction, its disruption and its rebuilding. Findings are examined in light of Turnbull's astute description and analysis of indigenous mapping techniques (Turnbull, 1994, 2000), Lynch's classical study of knowing the city (Lynch, 1960) and Vertesi's study of the importance of paying attention to user's meanings of maps (Vertesi, 2008) as well as mapping for community rebuilding in the international disaster literature (Chen et al, 2006; Diaz and Dayal, 2008). The research explores the challenges of translating mental or cognitive maps and visual knowledge onto paper when researcher and participant may not share visual cultures or may use and understand the same visual representations differently. It also explores how everyday spatial perception and real-life lived boundaries are raced and gendered by restrictions of navigation. Techniques are mental/cognitive map creation, map-elicitation through sketching, in-depth interviews and participant observation.

Visual methods and memory

Increasing attention is being given to the use of visual information in social research, providing a range of practical and theoretical frameworks for its incorporation and examination as more than illustration (Harper, 2002; Knowles and Sweetman, 2004; Pink et al, 2004; Rose, 2007). Literature addressing memory and landscape gives little attention to everyday knowledge in support of daily life, focusing rather on such things as mythological and personal history meanings of landscape (Comaroff and Comaroff, 1992; Santos-Granero, 1998) and cultural narrative as identity memory (Ebron,1998). Mapping techniques have been used in Taiwan as steps in a community-based disaster management programme to collect disaster experiences and empower participants (Chen et al, 2006). A Red Cross model for community-based psycho-social support based on a 'sense of place' suggests individual and group mapping to reveal local perspectives and knowledge (Diaz and Dayal, 2008). While Katrina accounts, written onto the landscape of survivors' memories, can be of mythic proportions, the focus here is on the disruption of the knowledge of daily living and how mapped accounts are connected to daily practice.

The field of science studies has long acknowledged the importance of visual, tacit and local knowledge in getting work done (Polanyi, 1967; Collins, 1974; Latour, 1986; Lynch and Woolgar, 1988; Knorr-Cetina, 1990; Ferguson, 1992; Jones and Galison, 1998; Vertesi, 2008). Gardner (2006) has documented multiple intelligence types, showing that people know things in ways beyond those quantified in verbal and mathematical intelligence tests. Specific to mapping practices, Mukerji (1984, 2006) has documented the cultural and political influences on historical western mapping conventions and conceptions of place while Turnbull has demonstrated the myriad ways indigenous peoples have mapped their worlds using all sorts of materials, method and style varying by culture (Turnbull, 1994, 2000). Content and form in maps as in art history, vary by what is culturally regarded as valuable 'to see' (Alpers, 1983; Latour, 1986). Such knowledge is also situational and indexical – knowledge in which component parts are interdependent including dependent on the situation (Suchman, 1987; Haraway, 1988; Henderson, 1999; Whalen et al, 2002). Less work has been done on the ties between tacit, visual, situated knowledge and identity.

Representational practice employs rendering conventions of the culture to which a person belongs, for example, deep one-point perspective in Renaissance Italy, bird's eye view in Egyptian tomb art and orthogonal projection in engineering drawings. Such rendering conventions colour the way in which the world of the person doing the rendering is or can be seen and conceptualized, whether in folk art, fine art, or technical art and in a range of levels of sophistication in articulation. Because there is never a clean one-to-one correlation between what is represented and what the representation means to the reader – whether a photograph, design drawing or hand-drawn map – the reader/maker brings a multitude of meanings to it.

Dialogue interactions around visual representations, termed 'photo-elicitation' or 'map elicitation' methods (Harper, 2002; Knowles and Sweetman, 2004; Pink et al, 2004; Rose, 2007) are particularly useful methodological tools because when images relevant to their lives are viewed by research participants they can work as reference and organizing devices for memory. A representation may help the viewer to remember things that took place in the setting, interactions with specific individuals, more details about the setting or its history, or any number of things not actually represented, but indexed to it for the viewer. Henderson (1999) documents how drawings serve such purposes in design engineering. When participants draw the map themselves memory helps organize the representation, its spatial particulars and its meaning. Capturing the individual's articulation of thoughts as they draw helps capture some of the memory process (Herlihy and Knapp, 2003; Vajjhala, 2005; Chen et al, 2006; Diaz and Dayal, 2008). Discussion around the drawing of maps of home neighbourhoods prior to relocation due to disaster can reveal salient spatial locations and relationships of daily life along with the manner in which those have been disrupted. Mapping and discussion of new residence locations in comparison to former locations has the potential to reveal the degree to which ruptures in situated identity have or have not been repaired. This study captures memory of daily life that has been disrupted and the mundane practices of contending with its recreation and adjustment through recording and naming. Like technical and scientific knowledge this knowledge too has conventions, preserved in the mind's eye of visual cognition and can be revealed to some degree employing dialogue around and through drawing maps.

Methods

In this study I asked Katrina survivor participants to draw maps of things most salient to them in the practice of everyday life before the hurricane, marking those that are no longer there. I also asked them to sketch important facets of their neighbourhoods of relocation to capture tacit, local, experiential knowledge. While some disaster recovery protocols incorporate community map making as instrumental in rebuilding devastated locales (Chen et al, 2006; Diaz and Dayal, 2008), little attention has been given to the *everyday knowledge* for navigating daily life: safe and unsafe areas; sources to obtain necessities; kinship and friendship networks; and community gathering spaces, among much more. I had built rapport in the community by the fact that I had previously brought student volunteers to help gut homes and later help build them so people were comfortable speaking with me candidly, plus they were anxious to talk about their experience. I was also able to pay them a small recompense for their time.

Relationships: Situated knowledge – *habitus*, race, class and gender

To address situated identity in transition it is important to pay attention to Bourdieu's (1977) *habitus*, experience-informed dispositions together with notions of practice that emphasize its situated nature and link to knowledge (Suchman, 1987; Haraway, 1988; Hutchins, 1995). Such syntheses help reveal embodied cultural distinctions of gender, race and class, without lapsing into essentialism. Inequality becomes 'real' in individuals' lived experience, shaping life choices and opportunities. Such meanings are not constant, but continually contested and negotiated. They can be and have been created, inhabited, transformed and destroyed. This research shows the process of *habitus* in transition and illustrates how it comes to be made up of 'bits and pieces, of partial evocations each apparently specific to its own level of perception but also partly evocative of some larger but again partial order' (Miller, 1994, p298). For disaster evacuees, connections, location, time and situated knowledge have been sorely disrupted. The intent, here, is to elucidate what these connections were previously, in order to understand a new situated knowledge as *habitus*, combined of new and old knowledge.

The messy practice of mind map elicitation

'Finger maps'

Katrina survivors expressed a need to convey spatial knowledge when talking about their former homes by making 'finger maps' on whatever space was available – chair arms, table tops, magazines. Two of the four grandmothers mentioned previously, describing the trolley line they used to take would trace it with a finger on the arm of their chair, filling in important points along the way such as the hospital where one of them had worked and the 'best bakery in New Orleans' where everyone bought French bread to make New Orleans famous Po' boy sandwiches. They also repeatedly used this technique to describe the route they took from their home to the superdome when they fled the floodwaters as well as the route's prevalence in their daily lives. A relocated Katrina survivor in her 40s used a similar approach, drawing with her finger on the table top to describe how her house was located in reference to other elements such as her mother's, now her sister's home, her church, and the beach, two blocks away, from which the water surged to flood the neighbourhood.

Visual and verbal ways of knowing

The idea to have Katrina survivors draw out on paper what they saw in their mind's eye and could translate into some sort of two-dimensional rendering came from the realization that people mapped their community in specific non-Cartesian ways, revealed by the failure to communicate through the Google map, mentioned above, in our encounter with Katrina survivors as user-centred design participants in the spring of 2006 (Graeber et al, 2006). If the Google map did not represent their knowledge of New Orleans, known locally by its historic wards, I wanted to know how they *did* see their home neighbourhoods spatially. So I started asking survivors to draw their own maps. However, this too met with mixed success, as described above, when some participants insisted they could not draw. These distinctions are about learned cultural conventions for rendering one's space. Turnbull's examples show Polynesian tied-reed navigation maps and Native American wood carvings of river meanderings, both resembling abstract art to the western eye, while Hutchins points out that in addition to using reed maps navigators immersed themselves in the ocean to distinguish currents corporally by temperature (Turnbull, 1994; Hutchins, 1995). Like the Katrina survivors who say they 'cannot draw', these visual cultures use other tacit ways of knowing.

Such distinctions are not necessarily limited to specific class, race or gender groups. Even within shared middle-class cultures of affluent computer-users, distinct preferences exist regarding the use of visual or verbal Mapquest or Google map features. Ask any electronic map user whether they look at the map or the verbal directions and an array of combinations from 'I never look at the map, I only use the verbal directions' to 'I only look at the map and not the directions because they confuse me' are forthcoming as well as combination approaches giving more or less weight to one or the other mode. One could argue that such preferences are based on visual or verbal proclivities; however, scholars have shown that socialization alters visual skills (Fausto-Sterling, 1997; Gardner, 2006;). These distinctions are about learned visual conventions and visual culture. In the following account a resilient young Katrina survivor in her 20s from New Orleans recounts how she navigated Houston, having arrived there after a 30-hour drive, exhausted and disoriented.

Map stories

Nicole,[1] native of New Orleans East

I meet Nicole at the Voodoo temple on Rampart where she is volunteering. An immaculately groomed young woman of colour with a trim figure, she works as an actress and model.

Verbal knowledge over Cartesian map knowledge

I: *Did you use maps in Houston?*

Nicole: '… I used Mapquest and followed the directions … When I print it out, I wouldn't even look at the map in Mapquest. I would just follow exactly what they say, take a right – … Once I got up there … (O my God, that was so horrible! I didn't realize – that 30 hour drive, just me and my cat. I didn't know where I was, that was terrible …) the apartment – I didn't leave it for five months until I decided to come home. I went the exact same places. That's all I did. I went to work … I was on a main street that now I can't remember the name of it. There was a grocery store; there was a Wal-Mart. There was a clothing store. So all I had to do was drive to the corner.'

Visual, situational knowledge, pre-disaster

'I love my city though … I know it by sight, thinking I know it constantly by heart. I remember that place and how to get there … somehow when I get in that area I recognize it by heart …'

Socialization to spatial knowledge

'Growing up, coming up … Your mother's going to tell you … 'Go out the door, make a right, go up four blocks, take a left.' You know, you might get to a main street by that time. And everything else pretty much intersects with the main street but there's so few main streets, it's easy, you learn it block by block as long as you know your right and your left … She'd be saying, 'Go up about three blocks, make a right. Go up two blocks, and I don't know the name of the street, but if you go up two blocks, you make a left.' …

I know it because I learned it. I think it's landmarks as well. I know my way around this city perfectly but I couldn't tell you half of the names of the streets.'

Local knowledge

Nicole: (giving directions to the temple) 'I would tell you, leave out the door, take a right, go straight down … this street until you get to Rampart, drive all the ways down, quite a ways. You're going to pass under the bridge, take a right on Rampart. I don't remember the name of the street … Where I'm sending you is where my grandparents grew up … So you would think that in 30 years I would know the name of some of these streets. What I'm saying is that I'm doing it block by block. I never learned them.'

Drawing what matters

Nicole, talking as she draws (see Figure 5.1):

'I was interested in the east … As soon as you get to the other side of the canal is houses. This is commercial and this side residential …

This used to be my favourite spot, once you pass these houses, there's a street here … there's a shell gas station, I know there's a service road – I don't know its name. The further you go *down here, and there's a pool hall where everybody's gone forever* (tapping on spot indicated). *Everybody who lives in the east goes to this place. That's how a lot of people were able to find out who was back in the city from this area, that I know of: what happened; who didn't make it. I started going there … [It's] not just neighbourhood. Everybody, not just east, goes there.* Sometimes they have territory problems … it's not like that. Everybody in the east goes there … That's where we all started going. They opened [when] I was in Houston for six years – six months – felt like six years …'

'I live right here on Lavern (pointing to former home location). I was in this house from four years old … It totally got trashed. No one in the area has returned at all … It's in the Seventh Ward, called "back of town". No one's back on that street at all and nothing's been done in that part …'

'Now, where I am living is still in the east … I thought I would be fixing up my house on Lavern. In this amount of time, everything should be done. There's no electricity, no grocery stores. I had to make the decisions that would be best for me. I was going to restore my house with my own money. But when I realized no one was back yet, I just keep working until I can move back …'

'But the area I live in, it's a complex and it's completely populated. Just that one complex; and it's police officers that live there. [points] And I feel safe in that area. I wouldn't feel safe living out there. I have a really great apartment and it's big, at a considerably reasonable price, because it's in the east …'

Nicole is relating that her knowledge of New Orleans is situational. Away from the city for six months and stressed over a potential identity theft, when she returns she gets confused because any one piece of navigational knowledge is related to others – remove one, and sense of location within space falls like a house of cards and must be rebuilt. She points out that for many the visual-navigation structure has been disrupted in so many ways, six months

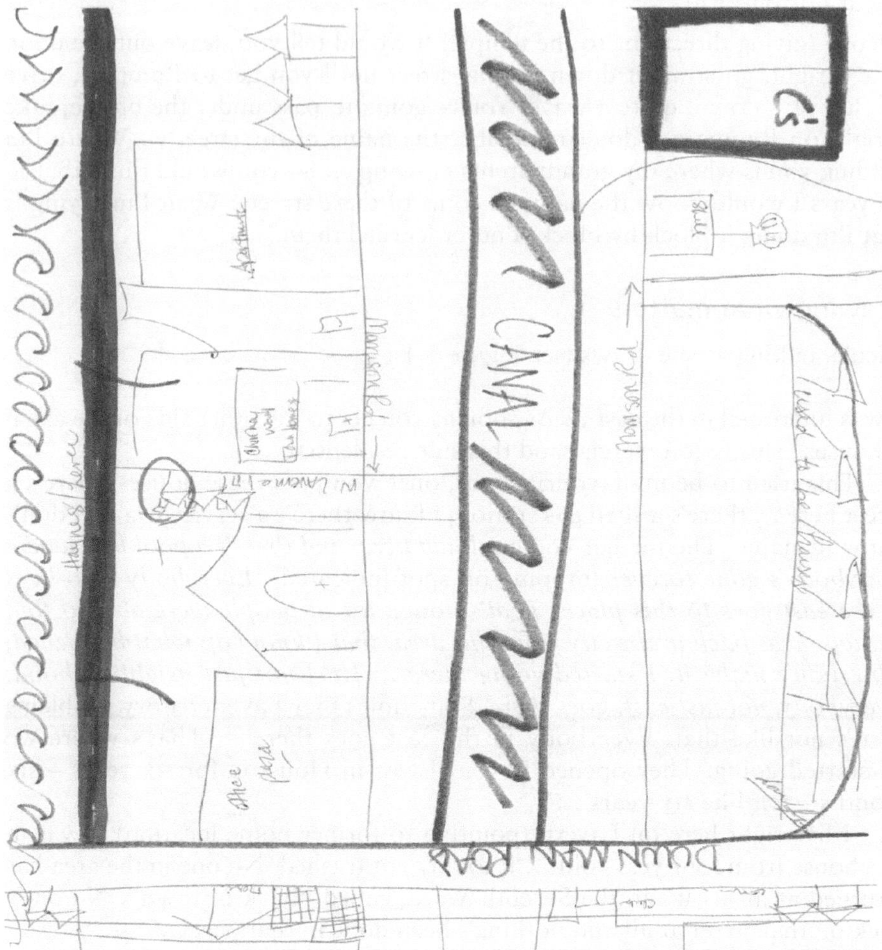

Map 1
Nicole's map of New Orleans East and the pool hall used as information center, located in the square marked 'C's'

Figure 5.1 *Map 1: Nicole's map*

in a different visual environment plus the trauma of everything hurricane-related. She works to rebuild and reconnect the pieces. Away in Houston and dependent on Mapquest she uses the verbal directions instead of the maps, again, because Cartesian maps are not the way she knows and navigates space. She explains how her family socialized her to know her neighbourhood block by block, and expresses a marked preference for directions that resemble that way of describing navigation. Moreover, similar to the desire for necessities in close proximity, expressed by the grandmothers in their house quest, described earlier, she creates her relocated world within tight distances and, once established, moves only within those familiar boundaries.

Like the Houston social workers, Nicole rejects the Google map of New Orleans that I show her because it does not represent her understanding of the landscape. In the map she draws she not only locates the spaces lost to her, her family home, her neighbourhood, but emphasizes the importance of a gathering point, a prime example of what Diaz and Dayal (2008) discuss as the re-establishment of 'sense of place': the pool hall which has become the most important place to her and others in its role as local information source and place to rebuild community connections – such a sacred place that turf wars are not allowed, a safe place, a centre where information circulates and radiates back into the community. The pool hall as place is exemplary of Diaz and Dayal's psychosocial support which helps people survive enormous personal losses following disaster through planning and coordinating responses to protect and improve their well-being instead of relying on outside help. In contrast, her family home, prominent in her mind map of the city, signifies loss of place since Nicole cannot return to rebuild it because she cannot feel safe in that unpopulated zone. Trying to at least be close to her old neighbourhood, she marks her new sense of place, her choice of an East side apartment where policemen reside, next door to a non-functioning church. Other populations feel differently about the police in New Orleans, yet also face the dilemma of a damaged mental map that used to highlight an array of safe places to sleep, given their old haunts have become unsafe since the hurricane.

Miss E.: Homeless advocate on the street

I run into Miss E. in the ladies room at Café Du Monde, in the French Quarter. I know Miss E., from our mutual volunteer work with Common Ground. A long-time homeless advocate, she herself, grew up on New Orleans' Streets as a teen runaway. I sit and talk with her and her friend at Miss E.'s Tarot reading table, set up adjacent to Jackson Square. As it gets later police cars arrive with lights flashing to chase out any homeless who are trying to sleep on the park benches. Two white female officers awaken a woman of colour who then begins vomiting. Miss E. helps the ill woman sit down, giving her some water and tissues. She banters with the officers in a friendly way. I interview her the following day at a sparsely furnished vintage residence hotel in the Quarter.

Asked if safe places for the homeless have changed since Katrina, Miss E. replies that the homeless are targeted more than ever and recounts witnessing a robbery near Jackson square as she draws it on her map.

Miss E: '... I was by myself, sleeping on top of Jackson Brewery (upstairs patio). And I had money in my pocket too because I'm a reader. They know that, they can target that. I woke up to that boy, "But that's my money, I worked hard all week for it, you're not gunno take that." And he was going to die before that man took his money and I was afraid when that happened, because when he knocked him down, he went down stair after stair after stair. I mean, there must have been 75 cement stairs that he went down and just a trickle of blood. And I don't know what happened, then, because I turned and got my stuff and went out the other way, because he turned toward me and I said, "You better not rob anybody else up here." I went back to my belongings to where my knife was, so I could protect myself and I went out another way ...'

'Most of the places in the Ninth Ward and the St Bernard Parish, that's where people were going to sleep. I had a young girl, she was 21. She's a bright young girl, but she was in the hospital three days last week. It was a mould infection, from sleeping in ungutted houses ...What's really changed ... it's that they're sleeping in abandoned houses. Actually most of the people in the Ninth Ward have family and they're all interconnected there ...'

Miss E. continues, mapping out details (see Figure 5.2)

'This is the wall, and a lot of us will sleep along inclines in doorways ... The wall is ... a feed line. It's up at Elysian Fields ... It's literally a wall, along the trolley ...'

'... Jackson Square is the meeting place for the homeless. That's where we end up at night ... Because of whoever it is trying to control the homeless, it's

Map 2
Miss E's map of safe and unsafe sleeping spaces for homeless persons

Figure 5.2 *Map 2: Miss E.'s map*

not safe anymore. In fact three girls were raped right here, [French Quarter] last Sunday. *Raped.* And I've been on and off these streets for a lot of years. And I'm worried about this square. It was *never* like this ...'

The homeless were together before Katrina, had some organization. Now the organizations that helped are shut down. Places with social workers or where a person could go to get a shave and get clean are gone. When people are resourceless they are more likely to do anything.

Miss E.'s map and narrative illustrate the mental shrinkage of safe places and resources from a homeless person's point of view, describing how former safe places have now become unsafe: the robbery she witnessed on second-story Jackson Plaza, the girls raped in the French Quarter. Locations for washing or where food that had been provided by institutions are also gone, as are abandoned buildings in the Ninth Ward that are now lethal with black mould. Another erasure is an 'ideal' abandoned warehouse boarded up by the city. Her political narratives are also connected to mapped space, a story of the woman politician who converted the benches around Jackson Square so the homeless could not lie down on them. The map too is about situated knowledge: places and spaces evoke details of habitual pattern such as what time the sprinklers come on in an otherwise good place to sleep, and events that render a place unusable: being arrested for sleeping on the levee in Algiers. The space becomes the story in her memory, a sort of time-line history of events that have taken place there, significant enough to render it safe or unsafe. Again we see place as prominent, but in this case it is the loss of place that continues even after homeless survivors return (Diaz and Dayal, 2008). And as more and more places are rendered unsafe by memory of dangerous incidents that have occurred there, whether perpetrated by the police or criminal elements, the mental map of safe places continues to shrink. [2] Unfortunately, it is not only the homeless that find safe spaces in New Orleans shrinking.

Gene: Barriers to community reformation (see Figure 5.3)

Gene, a 22-year-old African-American male, worked as he had before Katrina as a groundskeeper at a bed and breakfast where his mother was the housekeeper. Both returned from Dallas to their former jobs, about a year later. Gene, married, has a four-year-old son.

I: 'Who did you stay with when you came back?'

Gene: 'My auntie, same as my mom. That's where I always used to be, that's where I'll start. (begins drawing) ... That's Galvez (points). Now I'm going down like by the Lafitte Projects. See, that's like where we were staying, like all our life. But then the projects they still ain't open ... There's like a community centre, they cuttin' all the grass. They got a lot of police out there ... This is my grandma house, the next street over ... The projects are on Orleans.'

I: 'Did you have friends in the projects?'

Gene: 'Yes, that's where everybody I really know is at.'

I: 'Have you been able to see any of them since you got back?'

Gene: 'A couple of them I talked to when I was on MySpace, like on the computer, I got in touch with a lot of them on there. A lot of people I still ain't seen yet, ain't heard from ...'

I: 'When you were here, what were hang-out places for you and your crowd?'

Gene: '... the lake. The lake still ain't open ...I don't know the sides, you know. Cos I think they said like they got one side, you know, for blacks or somthin' like that, and one side – cos our side of the lake still ain't open yet, whichever side of the lake it is ...'

'See the park, that's where we used to take the children – it's up and running now ... the last time I took my little boy to get on the rides ... It's like the only park we got, ... like I was saying, at The Laffite we had a big old park and Sojourner Truth [community centre], that place there [pointing] we had a playground ... It's locked up ... There's swings and stuff there, but it's been under water.'

'We had a clinic right down on Orleans and Johnson [pointing]. It never opened.

And then I don't even know where another clinic's at. Oh, yeah, on Canal street. Some of them don't take Medicare or Medicaid or nothin.'

'When we come down to Claiborne, we got another corner store [Gene talks while drawing] ...

They got some gas stations and things that still ain't open down there. The school's still closed, the clinic, they's still closed ... McDonalds, all that's open.'

'I really didn't get out too much. Cos everybody was raised right there in the Lafitte housing apartments. Everybody I was knowing was like right there ...'

'The projects, here on Orleans, there's a whole lot of confusion on Dumont ... At the corner, where we at, the police come every day, harassing everybody – just want to cry. They need to open Lafitte back up. The project people are right there.'

I: 'Are you saying that police are harassing people who are on the street because they don't have a place to go?'

Gene: 'No, they got a place to go, they're just trying to stay close – to this project. They probably staying, w-a-a-y in Baton Rouge, somewhere, but they stay right here on Dumont just to be close to the project, right. I don't know, like everybody that stayed in Lafitte, like still, they keep comin' back. They can't leave. Like they gotta have Lafitte in eyesight, right ... And they stay

Map 3
Gene's map of the Lafitte Projects in the neighborhood of his aunt's house in the Treme, where he and his mother now live. Things missing were originally drawn in red, now a lighter gray.

Figure 5.3 *Map 3: Gene's map*

right in the area … cos that's where everybody from there, they know, hang out. That's their hang-out spot.'

I: 'And so they come back and they get harassed by the police, (**G:** 'Yeah') because they're on the street?'

Gene: 'Yeah, in large crowds, and all black male, yeah.'

I: 'They harass the men, but they come back anyway? (**G:** 'Um hum') even if the police have harassed them?'

Gene: 'Told 'em to leave. It might not be another day, might just be a couple hours later. I really don't sit outside unless my Mom, my Auntie are sittin' out with a chair. Cos it's crazy right now.'

I: 'Even if you're sitting outside your own house, they'll harass you? (**G:** 'Yeah') Even if you say, "I live here?"'

Gene: 'Yeah'

Gene's eyes flood while he tells me this. Gene and his fellows from the Lafitte projects in being denied access to their former homes and neighbourhood amenities are denied the function that the pool hall serves for Nicole and her New Orleans East compatriots.

Gene is emphasizing the need for the returning Katrina survivors to regain the situational space and place that is the location of their situated knowledge of the city. Expanding the idea of place as physical parameter in which human competence is achieved and a technique to offset the trauma of displacement (Diaz and Dayal, 2008), Gene and his former neighbours, locked out from the place that could potentially serve as hub of information sharing and community rebuilding are also being locked out from their cognitive orientation point to the city, a kind of compass rose to which all other points are related and contingent, indicated by Gene when he says, 'they gotta have Lafitte in eyesight.' This is not to say that people are lacking in navigation skills. It is to say that their comfort zone is located in a visual culture, a way of knowing space and place and navigating it that is familiar, like the narrow margins of navigation sought by the Grandmothers, Nicole, and Gene and his mother in their relocation spaces. To walk the city from the site of Lafitte is a piece of conditioned tacit knowledge. To navigate the city from another orientation point causes anomic disorientation, a loss of normative convention.

Obviously Katrina survivors have more extreme cases of anomie as every detail of their everyday lives has changed, even if they are fortunate enough to have been able to return to their former home and neighbourhood. To exclude them from the very orientation points they had as citizens of New Orleans increases the anomie. The Lafitte Projects were demolished even though they had very little water damage and UN experts spoke against the destruction as a violation of human rights. (UN News Center, 2008). See Allen and Maret (Chapter 7) for an account of the housing project controversy and the neighbourhood's cultural heritage. Also see Barrios (Chapter 6) for documentation of local voices that tried to speak up for public housing and were silenced by the way in which official representations of recovery planning excluded residents and facilitated the political and real estate interests that brought the housing projects down. A bit more context makes clear the larger implications of Gene's map account.

Quigley, the attorney leading the legal fight for public housing in New Orleans, pointed out that the residents of public housing in New Orleans are overwhelmingly African-American, female-headed households, often grandmothers on disability after a lifetime of working as a nurse's aide, cafeteria worker or minimum-wage job in the tourist industry. The destruction of public housing is a prime example that they are the ones who have been excluded from the work to rebuild the Gulf Coast. He states, 'It's very much a racial issue, it's very much a class issue, very much even a gender issue.'[3,4] On 28 March 2008 New Orleans Mayor, Ray Nagin, signed the demolition permit for Lafitte after a long delay which he said was due to the Department of Housing and Urban Development (HUD) not having handed over all the paperwork including documentation that the Housing Authority of New Orleans (HANO) had provided enough affordable housing for returning public-housing residents. On 31 March 2008, HUD Secretary Alphonso-Jackson, 'his tenure tarnished by allegations of political favouritism and a criminal investigation', announced his resignation 'amid the wreckage of the national housing crisis'.[5] Stating that he and the city council were comfortable that HUD was honouring its wishes, the mayor initiated the demolition of Lafitte, razing all but 196 units, temporarily preserved for returning public-housing residents. In the meantime, former project residents live in tents under the freeway that runs the length of the city and people like Gene's friends have no centre of orientation.

Loss due to disjuncture and dislocation

In conclusion, we find disjuncture and dislocation for almost every population coming to and going from New Orleans and the Gulf Coast post Katrina. Those of us outsiders who came in as researchers and volunteers held onto our AAA maps like security blankets in a world with no street signs. This researcher had a map annotated by a friend who had been down to volunteer, highlighting the specific landmarks that would help me navigate such as, 'take the second, not the first, Claiborne exit off the freeway', locations of places to stay at friends' homes and Common Ground volunteer centres, and one-way streets – an outsider's newly acquired, local knowledge, shared. Nevertheless our research team still had problems finding the freeway entrance with the signs down – we needed more extensive local knowledge than could be annotated on a map.

Katrina survivors attempted to navigate their new locations in Houston, Dallas and elsewhere by finding living accommodation where work and everyday supports such as grocery stores, discount stores, pharmacies and entertainment were in very close proximity. These strategies are exemplified by the Catholic grandmothers who finally accepted the offer of an unused former Black Baptist Church parsonage in walking distance to shopping and a community of support; Nicole, who set up her life within a few square blocks in Houston; and Gene's family who also set up their lives within blocks of work, schools and shopping in Dallas.

Upon return, survivors find that their situational navigational knowledge of their city is not the same as before, experiencing loss of place (Diaz and Dayal, 2008) even as they try to re-engage it. Nicole, despite the resource of the pool hall as neighbourhood gathering place, mapped her whole neighbourhood in New Orleans East as missing. The city landscape itself is changed, even if damaged landmarks are still in place. When some pieces of our cognitive maps are lost due to time and trauma, we can get lost in our own neighbourhoods because spatial navigation depends on contingent and inter-related situational knowledge. Like the rest of us who may not know the name of the street that we turn on to get to our best friend's house, though we recognize it when we see the garden of the house on the corner, Katrina survivors 'know' their route, situated in their neighbourhood. And in the neighbourhood some things are missing completely or partially, such as the hulk of a building still standing while the clinic that used to be there is gone, the 'black' side of the lakeside closed, or the home point for navigation locked and guarded by a belligerent police force.

Nicole, Miss E. and Gene map the gaps, the things that are missing. Other individuals who drew maps pointed out things crucial to their lives that they now must travel miles to obtain. The owners of the bed and breakfast where I stayed pointed out the lack of clinics, grocery stores, mechanics near the French Quarter and Fauberg Marigny. They drive either out to suburban Metairie or St. Bernard Parish, itself still suffering from 100 per cent destruction of its infrastructure, to buy food supplies and for car repairs, flat tyres being a frequent occurrence, given the city's simultaneous demolition and reconstruction.

Found: Community and social supports

Despite these great losses, the rendering of cognitive maps also reveals things found by both hurricane survivors who have decided to stay in Texas and those who have elected to return to New Orleans. Kinship and other networks have remained intact. People re-establish their 'sense of place' (Daiz and Dayal, 2008) by taking action. They return to New Orleans for rites of passage: birthdays, funerals, weddings, Mardi Gras or to participate in community activities such as providing weekly music for the Jazz Mass at historic St. Augustine Church. Sketched mental maps document that pre-disaster people lived in close proximity to their churches, schools, work and kinship networks and that, despite hardships of closed schools, parks and churches, survivors still maintain contact with these networks just as tsunami survivors in Wuring returned to their fishing village despite the advice of outside experts (Boen and Jigasu, 2005). Some homeless people are part of the homeless activist network and privately run, NGO support structures. While the homeless population is larger and support structures are stretched thinner than ever before, knowledge of the remaining resources was still extant and conveyed among the homeless community Miss E. described, though this interview took place prior to the more recent growth of the tent city of returnees camped under the freeway.

Some implications

Spatial and navigational knowledge is tacit, situational and indexical. Modes of spatial representation and understanding are not necessarily Cartesian, nor universal, but are influenced by the culture in which navigation is learned and practised. Such navigational knowledge is spatially situated and interrelated with memory and other spatial knowledge, so it can be disturbed by prolonged dislocation in which the connections between nodes of knowledge may become undone due to lack of exposure to points of daily reinforcement, including, though not limited to, viewing landmarks, salient to the viewer, which index other visual knowledge for navigating the city.

Representation and mental maps are indexed to different types and forms of personal memory, such as: childhood experience and socialization, significant cultural events such as Super Sunday and Mardi Gras, traumatic events such as evacuation in the wake of hurricane, tsunami or earthquake and everyday life routines such as shopping for groceries or fishing for livelihood. It is the linkages between these myriad nodes of knowledge along with the nodes, themselves, that make up situated knowledge. Using cognitive map elicitation by asking participants to draw the salient features of their local terrain elicits an interaction between the researcher and the participant that can be valuable to both in terms of understanding why disaster survivors may feel disoriented even when they return to their neighbourhoods.

The research presented here not only illustrates the rich details of everyday knowledge that sketching cognitive maps can reveal. It also illustrates important different ways of knowing a locale than Cartesian and GPS maps can reveal and a method for documenting such contextual and multiple meanings of space and place. An unplanned side benefit of the process was that it allowed participant map makers to recognize that their spatial disorientation was not a flaw in themselves but another aspect of post-disaster trauma.

Policy implications

The United Nations experts' critique of the destruction of public housing and their mention of the lack of consultation with those affected could have been addressed by just such participant sketched mapping as described here (UN News Center, 2008). This technique has been used in applied anthropology and geology to understand members' meanings of space and place (Nietschmann, 1995; Eghenter, 2000; Chapin, Lamb, and Threlkeld, 2005; Herlihy and Knapp, 2003), in engineering and public policy (Vajjhala, 2005) and in international post disaster protocols (Diaz and Dayal, 2008). Case studies from Erickson's (1976) documentation of the social conflict created by the random housing arrangements for people displaced in the wake of the Appalachian Buffalo Creek flood to Boen and Jigyasu's (2005) analysis of the importance of cultural considerations to ensure the sustainability of interventions in disaster reconstruction after the Indonesian earthquake, Indian Ocean tsunami, Bam earthquake and Gujarat earthquake point out that had attention been paid to the culture and input of communities people could have been resettled in ways that would have aided their recovery and saved expense as well. In contrast, Chen et al's (2006) study of an integrated community-based disaster management programme to encourage local people to take action in hazard mitigation in the wake of earthquakes and typhoons in Taiwan points out the strength of community involvement both in terms of more practical, long-term outcomes and empowerment. That project also included mapping disaster experiences. Picou (2000) documents that even five years after the Exxon Valdez oil spill in Alaska personal and community healing could take place through ceremonial interaction based on indigenous cultural traditions to transform chronic disaster impacts. Fussell (2006) points out another missed opportunity in the wake of Katrina: given the ideology of disaster management, social services in shelters disempowered the people they were supposed to help by making them dependent on aid rather than supporting them in finding their pre-existing social networks, which could have been extremely helpful to them. Once again, policy makers ignored the desire of a community to reform itself, instead arbitrarily tearing it asunder by prohibiting residents to return to their community of origin, based on the hopes of redevelopers that their plan to mix strangers together is better. All of which might raise some questions, despite Mayor Nagin's confidence, about how many and which local voices, which the UN report suggested should be heard, had input about the redevelopment plans for Lafitte and other projects instead of the noise of bulldozers. Mapping techniques as described here would have gone a long way toward including those voices in an accessible yet complex depth for planners who cared about the city's diverse population's input.

Notes

1 Pseudonyms are used throughout.
2 See Allan and Maret (Chapter 7) on the doubling of the number of homeless since the storm and the interrelationship between public housing and the working poor.
3 www.democracynow.org/2007/12/12/hud_poised_to_tear_down_new.
4 On 28 February 2008, United Nations experts on housing and minority rights called for the protection of the human rights of African-Americans affected by Hurricane Katrina and to halt the demolition of public housing, stating, 'the demolition of public housing, in combination with the spiralling costs of private housing and rental units, are helping to drive people, primarily African-Americans, into destitution' (UN News Center, 28 February 2008).
5 Associated Press, 31 March 2008, www.nola.com/newflash/index.ssf?/base/politic-14/1206965343134080.xml&stor.

References

Alpers, S. (1983) *The Art of Describing: Dutch Art in the Seventeenth Century*, University of Chicago Press, Chicago, IL

Associated Press (2008) 'HUD secretary Jackson resigns amid probe', 31 March, www.nola.com/newflash/index.ssf?/base/politic-14/1206965343134080.xml&stor, accessed 31 March 2008

Bourdieu, P. (1977) *Outline of a Theory of Practice*, Cambridge University Press, Cambridge, MA

Bourdieu, P. (1990) *The Logic of Practice*, Polity, Cambridge, MA

Boen, T. and Jigyasu, R. (2005) 'Cultural considerations for post disaster reconstruction post-tsunami challenges', paper presented at the UNDP Conference, September, Beijing http://www.adpc.net/IRC06/2005/4-6/TBindo1.pdf

Chapin, M., Lamb, Z. and Threlkeld, B. (2005) 'Mapping indigenous lands', *Annual Review of Anthropology*, vol 34, pp619–638

Chen, L. C., Liu, Y. C. and Chan, K. C. (2006) 'Integrated community-based disaster management program in Taiwan: A case study of Shang-An village', *Natural Hazards*, vol 37, pp209–223

Collins, H. (1974) 'The TEA Set: Tacit knowledge and scientific networks', *Science Studies*, vol 4, pp165–186

Comaroff, J. and Comaroff, J. (1992) *Ethnography and the Historical Imagination*, Westview, Boulder, CO

Democracy Now! (2007) 'HUD poised to tear down New Orleans public housing developments', 12 December, www.democracynow.org/2007/12/12/hud_poised_to_tear_down_new, accessed 9 April 2008

Diaz, J. O. P. and Dayal, A. (2008) 'Sense of place: A model for community based psychosocial support programs', *Australasian Journal of Disaster and Trauma Studies*, vol 1, pp1–12

Ebron, P. (1998) 'Enchanted memories of regional difference in African American culture', *American Anthropologist*, vol 100, no 1, pp94–10

Eghenter, C. (2000) 'Mapping peoples' forests: The role of mapping in planning community-based management of conservation areas in Indonesia', Washington DC Biodiversity Support Program

Erickson, K. (1976) *Everything in Its Path: Destruction of Community in the Buffalo Creek Flood*, Simon & Schuster, New York

Ferguson, E. (1992) *Engineering and the Mind's Eye*, MIT Press, Cambridge, MA

Fausto-Sterling, A. (1992) *Myths of Gender: Biological Theories about Women and Men*, 2nd ed, Basic Books New York

Fussell, E. (2006) 'Leaving New Orleans: Social Stratification, Networks, and Hurricane Evacuation', in *Understanding Katrina: Perspectives from the Social Sciences*, sponsored by Social Science Research Council (SSRC), understandingkatrina.ssrc.org/Fussell, accessed 7 April 2008

Gardner, H. (2006) *The Development and Education of the Mind: The Selected Works of Howard Gardner*, Routledge, London

Graeber, R., Kerne, A. and Henderson, M. K. (2006) 'ZooMICSS: A zoomable map image collection sensemaking system (The Katrina Rita context)', in Proceedings of the 14th annual ACM international conference on Multimedia, ACM, New York, pp795–796

Haraway, D. (1988) 'Situated Knowledges: The science question in feminism and the privilege of partial perspective', *Feminist Studies*, vol 14, no 3, pp575–600

Harper, D. (2002) 'Talking about pictures: A case for photo-elicitation', *Visual Studies*, vol 17, no 1, pp13–26

Henderson, K. (1999) *On Line and On Paper: Visual Representations, Visual Culture, and Computer Graphics in Design Engineering*, MIT Press, Cambridge, MA

Herlihy, P. and Knapp, G. (2003) 'Maps of, by, and for the Peoples of Latin America', *Human Organization*, vol 62, no 4, pp303–314

Hutchins, E. (1995) *Cognition in the Wild*, CMIT Press, Cambridge, MA

Jones, C. and Galison, P. (eds) (1998) *Picturing Science Producing Art*, Routledge, London and New York

Knorr-Cetina, K. (1990) 'Image dissection in natural scientific inquiry', *Science, Technology and Human Values*, vol 15, pp259–283

Knowles, C. and Sweetman, P. (2004) *Picturing the Social Landscape: Visual Methods and the Sociological Imagination*, Routledge, New York

Latour, B. (1986) 'Visualization and cognition: Thinking with eyes and hands', *Knowledge and Society: Studies in the Sociology of Culture Past and Present*, vol 6, pp1–40

Lynch, K. (1960) *The Image of the City*, The Technology Press and Harvard University Press, Cambridge, MA

Lynch, M. and Woolgar, S. (eds) (1988) 'Special Issue on Representation in Scientific Practice', *Human Studies*, vol 11, pp2–3

Miller, D. (1994) *Modernity – An Ethnographic Approach: Dualism and Mass Consumption in Trinidad*, Berg, Providence, RI

Mukerji, C. (1984) 'Visual language in science and the exercise of power: The case of cartography in early modern Europe', *Studies in Visual Anthropology*, vol 10, pp30–35

Mukerji, C. (2006) 'Printing, cartography and conceptions of place in Renaissance Europe', *Media, Culture and Society*, vol 28, no 5, pp651–669

Nietschmann, B. (1995) 'Defending the Miskito Reefs with maps and GPS: Mapping with sail, scuba, and satellite', *Cultural Survival Quarterly*, vol 18, no 4, pp34–37

Picou, S. J. (2000) 'The "talking circle" as sociological practice: Cultural transformation of chronic disaster impacts', *Sociological Practice: A Journal of Clinical and Applied Sociology*, vol 2, no 2, pp77–97

Pink, S., Afonso, A. I. and Kürti, L. (eds) (2004) *Working Images: Visual Research and Representation in Ethnography*, Routledge, New York

Polanyi, M. (1967) *Personal Knowledge*, Routledge and Kegan Paul, London

Rose, G. (2007) *Visual Methodologies*, 2nd ed. Sage, Thousand Oaks, CA

Santos-Granero, F. (1998) 'Writing history into the landscape: Space, myth, and ritual in contemporary Amazonia', *American Ethnologist*, vol 25, no 2, pp128–148

Suchman, L. (1987) *Plans and Situated Action: The Problem of Human–Machine Communication,* Cambridge University Press, Cambridge

Turnbull, D. (1994) *Maps are Territories: Science is an Atlas*, University of Chicago Press, Chicago, IL

Turnbull, D. (2000) *Masons, Tricksters and Cartographers: Comparative Studies in the Sociology of Scientific and Indigenous Knowledge*, Taylor and Francis, New York

UN News Centre (2008) 'UN experts call for protection of housing rights of Hurricane Katrina victims', 28 February, www.un.org/apps/news/story.asp?NewsID=25782&Cr=housing&Cr1, accessed 31 March 2008

Vajjhala, S. (2005) 'Mapping Alternatives: Facilitating Citizen Participation in Development Planning and Environmental Decision Making', PHD thesis in Engineering and Public Policy, Carnegie Mellon University, Pittsburg, PA

Vertesi, J. (2008) 'Mind the gap: The London underground map and users' representations of urban space', *Social Studies of Science*, vol 38, no 1, pp7–33

Whalen, J., Whalen, M. and Henderson, K. (2002) 'Toward a naturalistic sociology of work: Improvisational choreography in a teleservice job', *British Journal of Sociology*, vol 53, no 2, pp239–258

6
Post-Katrina Neighbourhood Recovery Planning in New Orleans

Roberto E. Barrios

Anthropological perspectives on recovery planning

Disasters result from human practices that enhance the destructive effects of geophysical phenomena and place certain sectors of an affected population at greater risk of harm during catastrophic events (Blaikie et al, 1994; Oliver-Smith, 1999; Bankoff and Hilhorst, 2004). Because disasters are never purely 'natural' but are the result of relationships between people and environment, social scientists see disaster reconstruction as a critical moment when policy makers and disaster-affected populations may reflect on and address the human practices that give disasters their form and magnitude (Wisner, 1995; Oliver-Smith, 1999; Ganapati and Ganapati, 2009). These reflections take place when people ask the question: What should be reconstructed?

One challenge for contemporary state societies is reconciling the technical languages of economics, risk assessment and modernist planning with the location-specific ways people produce and experience urban space. As a number of anthropological studies have shown (Holston, 1989; Douglas, 1992; Rabinow, 1995; Scott, 1999; Mitchell, 2002; Caldeira and Holston, 2005), expert planners and policy makers often credit these technical languages as being a culturally neutral and objective means of describing the world and its problems. Nevertheless, these same studies have shown that economics, risk assessment and modernist planning are specific to the cultural, political and environmental settings in which they were devised. These expert languages feature their own cultural and material biases in how they establish parameters for defining social problems, people's experience of material and social worlds, and the relationships people should have towards one another and their environments to produce social well-being. Most importantly, these studies have also shown how the failure to recognize the cultural and environmental histories and particularities of expert knowledge, when combined with its rigid application, can create conditions of social marginalization and environmental degradation (Escobar, 1995; Bankoff and Hilhorst, 2004; Barrios, 2009). Consequently, a number of disaster researchers (Wisner, 1995; Shaw et al, 2003; Ganapati and Ganapati, 2009; Tanaka et al, 2009) have emphasized the importance of participatory planning as a means of adapting expert knowledge to the social and environmental particularities of disaster-affected localities.

In post-Katrina New Orleans, the United States Congress required that recovery plans for the city feature a participatory component that included residents. Early planning initiatives like Mayor C. Ray Nagin's 'Bring New Orleans Back' plan (autumn 2005–spring 2006) met widespread criticisms from city residents for lacking a participatory procedural structure (Davis, 2006). City officials, planners and representatives of national philanthropic organizations have represented subsequent planning processes [like the City Council-Lambert Plan (summer 2006) and Unified New Orleans Planning (autumn 2006–spring 2007)] as mechanisms of shared governance through which city residents may stipulate the city's reconstruction directive to expert planners (Warner, 2006, 2007). In the words of New Orleans city council members Shelly Midura and Arnie Fielkow, the meetings between city residents and expert planners that comprised these two latter planning initiatives were 'a process that is bottom–up, and not top–down' that was 'very important in order to indicate what you [residents] want the directive to the planners to be' as the New Orleans city council tried 'to put together a single plan' for the city's reconstruction (Neighbourhoods Planning Network meeting, 7 June 2006).

These comments by New Orleans city council members articulate a common representation of official neighbourhood recovery planning in post-Katrina New Orleans. This representation portrays experts (architects, city planners) and the planning documents they produce as value-neutral agents that divulge the ideas of city residents to a wider audience. In such a representation, the contribution on the part of planning experts is a technical one, and city residents are provided with a mechanism for communicating with a broader audience. Planning is thus depicted as a rational scientific enterprise, where epistemological differences – the conditions of possibility for making statements about the nature of social well-being and the meaning of disaster recovery – between experts and residents are not thought to complicate the process.

The anthropological literature on planning, however, paints a more complex picture. In the case of urban planning, participatory processes can be complicated by the epistemological assumptions on the part of expert planners on how urban spaces should be structured and used (Holston, 1989; Rabinow, 1995; Calderia and Holston, 2005). These epistemological assumptions are a reflection of the cultural, political and environmental specificity of expert knowledge on the nature of social well-being and the practices necessary to produce it. The works of Teresa Caldeira and James Holston (2005), Paul Rabinow (1995) and James C. Scott (1999) have shown how, in the case of urban planning in 20th-century nation states, government planners have often believed that certain architectonic arrangements of urban space could produce predictable conditions of social modernization. Social modernization is a concept that has varying meanings in different social contexts, but has historically been associated with attempts to eradicate stark socio-economic differences, produce biologically optimized bodies, create capitalist 'free' markets and extensive mechanisms of state surveillance. These goals, however, were never met in practice. In some cases, utopian visions of social modernization were subverted by the very populations destined to live in the urban spaces engineered by expert planners as people carved out niches and created enclaves that reflected their distinct ways of structuring and experiencing urban space. In other instances, modern urbanism inadvertently created those very social conditions it set out to eradicate (like social inequity and marginalization) through the dissection and privatization of public space.

Over the past three decades, the epistemological assumptions of modernist planning have been the subject of extensive critiques by social scientists, leading to a reconsideration of high modernism as a suitable approach to urban planning. But even as modern urbanism has waned in its status as a universally applicable solution to social challenges, social scientists have seen the emergence of neoliberalism as a privileged point of reflection when considering issues of urban development and disaster reconstruction (Ong, 2006; di Leonardo, 2008; Schuller, 2008). Neoliberalism, in this case, is a term used to describe the idea that market liberalization (the privatization of public services, the use of public resources to encourage capital investment, the reduction of labour, environmental and fiscal regulations) will lead to optimal social ends, and the notion that all domains of human experience can be reduced to cost–benefit analyses and capitalist utility. Although many policy-makers often herald neoliberalism as a ready-made solution to social challenges, critical theorists have warned that neoliberalism threatens to exacerbate socio-economic inequities and enhance conditions of environmental degradation, two root causes of disasters (di Leonardo, 2008; Schuller, 2008).

An approach that highlights the epistemological differences between expert knowledge and the ways disaster-affected populations structure and experience urban space stands to make an important contribution to the literature on participatory disaster recovery planning. Studies around the globe have already noted a number of challenges complicating the execution of participatory post-disaster reconstruction. In the case of South Africa, Ben Wisner (1995) has noted that historically marginalized populations can develop a profound mistrust of expert planners as they associate these professionals with the local and national governments that have historically underserved, exploited or discriminated against them. In the case of Turkey's 1999 earthquake, Emel and Sukumar Ganapati (2009) also noted how the narrow definition of 'stakeholder' in recovery planning processes by international agencies like the World Bank can exclude important voices (like local governments and civil society organizations), and how an inflexible institutional emphasis on financial cost–benefit can impede the adaptation of reconstruction programmes to the cultural particularities of an affected site. At the same time, works like Tahiro Tanaka et al's (2009) research on the aftermath of the 1995 Kobe earthquake have emphasized the importance of the use of GIS (geographic information systems) in community charrettes as a means of combining expert and local knowledge. While these and other works have highlighted critical challenges of participatory recovery planning, the exploration of epistemological differences between planning experts and disaster-affected populations remains largely under explored in the disaster reconstruction literature. In this chapter I argue that understanding the dramatically different ways expert planners and disaster-affected populations define social well-being and experience urban space is an important first step towards understanding the things that disaster survivors articulate during participatory planning processes.

Over the course of this research, I used three questions to structure my observations of recovery planning activities: (1) What kinds of ideas about the use of urban space, the nature of neighbourhood and community, and the practices necessary to produce social well-being will the different actors involved in neighbourhood recovery planning (architects, residents, local government officials, non- and for-profit developers) bring to planning encounters? (2) How will critical differences in the ways these actors conceptualize neighbourhood and social well-being be negotiated in the creation of recovery plans? And finally, (3) what are the social implications of the ways different visions of neighbourhood recovery are negotiated in official post-Katrina recovery planning?

My research has focused on New Orleans' Fourth Planning District, a central area of the city that is contiguous to the Central Business District and the French Quarter, the city's financial and tourism centres. Beginning in the summer of 2006 and continuing into the autumn of 2007, I conducted observations of the City Council-Lambert planning meetings and the Unified New Orleans Planning (UNOP) process. I also conducted observations at the meetings and activities of the Neighbourhoods Planning Network (NPN), the St. Augustine Church Council (a group of residents and church officials organized to revitalize this historic church), and the Fauburg Tremé Neighbourhood Association. I spoke with architects, city planners, civil engineers, neighbourhood association officers and members, and activists who live or work in the Fourth Planning District. I also collected planning documents that are indicative of the kinds of ideas being contemplated for the city's reconstruction by official planning teams.

The social production and meanings of urban space in New Orleans

Since the founding of New Orleans in the early 18th century, residents have structured the city's space as a means of producing racialized class and ethnic differences. During the city's early years, land ownership in the area that is today's French Quarter was limited to people who could claim full French genealogies (Johnson, 1992; Campanella, 2006). Consequently, people of African ancestry and the descendants of unions between African, Native American and European residents found themselves limited in the places they could reside. Eventually, the first expansions beyond the colonial city, like the neighbourhood of Tremé, became places of residence for recent European arrivals, African-Americans who earned their way out of bondage, exiles from Saint Domingue, and an emerging group of francophone New Orleanians whose genealogies featured European, Native American and African ancestors (Campanella, 2006; Dowdy, 2008). This latter socially differentiated group would eventually come to self identify as Creole in the late 19th century. As Allen and Maret (Chapter 7) show, Tremé was a highly integrated and socially diverse place during its early years, but the demographic profile of the neighbourhood changed towards one of racial polarization in the following two centuries.

The end of the American Civil War, the emancipation of African-American slaves and the racial politics of Reconstruction had a profound effect on the urban landscape of New Orleans. Following the Civil War, New Orleans witnessed a rural–urban migration of emancipated African-Americans. For many francophone New Orleanians who self identified as Creoles, the presence of emancipated slaves in the city threatened their established social position. Under 18th-century colonial law, Native American, Creole and African-American residents were encouraged to become 'more French' through the adoption of colonial language and culture (Johnson, 1992). Consequently, many Creoles (who, in the Anglo-American dichotomous model of black and white, are often categorized as black) developed a separate identity from the African-American populations who remained in bondage through the mid-19th century. In the context of post-Civil War Reconstruction, Anglo-American dichotomous discourses on race threatened to erase this important social distinction, and many creoles created new residential enclaves toward the city's northeast as a means of differentiating themselves from African-Americans (Campanella, 2006).

Following desegregation in the early 1960s, New Orleans lost 200,000 residents, many of these being New Orleanians who self identified as white, and who migrated either westward toward the emerging suburban city of Metairie, or eastward to St. Bernard Parish. Metairie eventually became known for the racial profiling practices of its long-time sheriff, Harry Lee, and the electoral support on the part of many of its residents of former Ku Klux Klan Grand Wizard David Duke in his race for the United States House of Representatives (Sorant et al, 1984; Nossiter, 2007).

In New Orleans, African-American populations who have historically experienced attempts at curtailing their geographic and social mobility have developed culturally-particular ways of structuring and experiencing urban space and public participation (Breunlin and Regis, 2006; Lipsitz, 2006; Irazábal and Neville, 2007). Lipsitz (2006), for example, has suggested that working class African-American populations, when faced with limited prospects for geographic mobility, develop profound attachments to place that differ from populations who have not experienced spatial forms of discrimination. In New Orleans, one of the ways in which working class African-Americans structure and develop attachments to the urban spaces they occupy is through ritual parading practices known as Second Lines and Mardi Gras Indian Super Sunday. Second Lines are pedestrian parades that feature a jazz quintet (trombone, tuba, trumpet, snare drum and base drum), a group of uniformed dancers (first line) and a group of participating spectators (second

Source: Photograph by author

Figure 6.1 *Peter Parker and Gerald Platenburg of the Nine Times Social Aid and Pleasure Club in New Orleans' Seventh Ward*

line) (Figure 6.1). Mardi Gras Indian Super Sunday, in turn, takes place in the weekends surrounding St. Joseph's Day (19 March), and features pedestrian parades of organizations called Mardi Gras Indian Tribes whose members dress in elaborately beaded and feathered costumes aesthetically inspired by Native American Great Plains tribes (Figure 6.2). Both these practices are dated to the late 19th century, the era of American state-sanctioned segregation, but are considered to be part of broader and older African diasporic parading practices that extend into the Caribbean (Regis, 1999; Breunlin and Lewis, 2009).

Second Lines and Super Sunday are the ritual counterparts of quotidian socialization practices that comprise the experiences through which the durable, transposable and embodied dispositions (Bourdieu, 1977) of many working class New Orleanians are shaped. As Henderson (Chapter 5) shows, the dispositions of Tremé residents are shaped by histories of experiencing 'extended kinship networks; networks which included kin and tight friendships, centres for socializing and exchanging information, and clerks in commercial chains that treated a person like a neighbour.' Moreover, Henderson notes that 'such situated, local knowledge is its own way of knowing. It is tacit, site specific, and not captured in the Cartesian planes of commercial maps and GPS coordinates.' Consequently, many New Orleans residents involved in recovery planning initiatives conceptualize recovery in terms of the reinstitution of the city's pre-Katrina population, which was responsible for

Source: Photograph by author

Figure 6.2 *Mardi Gras Indians parade in Uptown, New Orleans*

the continuation of quotidian and ritual practices that shaped the dispositions of present and future generations.

The Fourth Planning District

The city of New Orleans is officially divided into 73 neighbourhoods that are grouped into 13 planning districts. Despite its central location, many of the Fourth Planning District's 79,000 residents faced significant challenges before the storm, having a low percentage of home ownership (23 per cent, compared to a 41 per cent city average) and a median income of 17,930 US dollars (compared to 31,200 US dollars citywide median) (US Census Bureau, 2000; UNOP, 2007). Prior to Katrina, 83 per cent of the district's population self-identified as African-American, 13 per cent as white, and the predominant remainder as Hispanic in the 2000 US Census (US Census Bureau 2000; Campanella, 2006; UNOP 2007).

The Fourth Planning District featured several public housing projects, including Lafitte, Iberville, St. Bernard, B.W. Cooper, three of which were slated for demolition after the storm by order of the Housing Authority of New Orleans (HANO) and the Department of Housing and Urban Development (HUD). Laffite public housing is located in the vicinity of Tremé, which has witnessed contestations between developers, city planners, residents and activists over the use of urban space during the last four decades. As Allen and Maret (Chapter 7) demonstrate, urban renewal projects during the 1960s had profound deleterious impacts on the neighbourhood that persist in the memory of many residents and serve as a point of reference for their interpretation of proposed recovery plans. In the 1960s the neighbourhood was divided by the construction of an elevated interstate highway (I-10) over N. Claiborne Avenue, a focal area of economic and social activity. The interstate's construction socially fragmented the neighbourhood and overshadowed local businesses (Sorant et al, 1984; Lacho and Fox, 2001).

In the 1970s the neighbourhood was the focus of a disputed urban development project that displaced approximately 175 families and resulted in the creation of Louis Armstrong Park, an allegedly public space whose fences curtailed the movement of neighbourhood residents through its grounds (CUPA, 1995). In the last 20 years, Tremé has seen the growth of a new, more affluent, resident faction that has purchased real estate in the neighbourhood's north-eastern border, which runs along Esplanade Avenue. These residents have set out to socio-economically transform the neighbourhood, but the gentrifying tendencies of their practices have created tensions with other grass-roots organizers and residents (New Orleans Renovation, 2006). The Fourth Planning District, then, is a socio-politically complex locality where neighbourhood real estate investors, federal and city housing agencies, local government officials, grass-roots activists and concerned residents have different (and sometimes conflicting) ideas about the meaning of neighbourhood well-being and recovery.

The City Council-Lambert Planning Process

Prior to Katrina, there was no established city-wide master plan for New Orleans. The city Planning Commission had worked on a plan since 1992, but had not completed it prior to the hurricane. After the storm, the Mayor's office created the 'Bring New Orleans Back' (BNOB) commission, an expert panel that released a preliminary plan for the city's reconstruction on 11 January 2006 (Davis, 2006). The preliminary plan met with significant resistance from city residents because it lacked open participation and de-prioritized the reconstruction of heavily flooded neighbourhoods. As a response to the controversial BNOB plan, the New Orleans City Council organized a supplemental planning process focused on neighbourhood recovery that would be incorporated into the city's unfinished Master Plan (Gerber, 2006). In the spring of 2006, the New Orleans city council hired architects Paul Lambert and Bernard Zyscovich to assist them with this task. Although the New Orleans City Planning Commission's chief planner represented this process as a participatory endeavour over the course of ethnographic interviews, Tremé residents were not consulted during the plan's initial drafting stages.

On 13 July 2006, Bernard Zyscovich presented a preliminary version of the City Council-Lambert plan for District 4 to Tremé residents at St. Augustine Church, one of the neighbourhood's principal landmarks. As Bernard introduced himself, he indicated that his firm was based in South Florida, which made a Tremé resident sitting behind me whisper in disbelief: 'Did he say *Miami*!?'

The Zyscovich plan focused on the architectonic transformation of one of the neighbourhood's thoroughfares, North Claiborne Avenue, with the intention of economically revitalizing the area. Bernard spoke: 'We're going to bring up a pretty radical idea. We want to re-establish the historic Claiborne Road with a boulevard, sidewalks, and tree planting. We are looking at the idea of bringing down parts of I-10. The idea is to bring people down into a beautiful boulevard' (Zyscovich presentation, St. Augustine Church, 13 July 2006).

Behind Bernard, a portable projector displayed a watercolour-like image that showed a wide boulevard with young trees planted on sidewalks and what seemed like boutique retail spaces. The radical idea was the removal of a stretch of high-rise interstate highway (I-10) that dissected Tremé during the late 1960s. The highway was constructed over Claiborne Avenue, which was once a principal site of economic activity. The interstate's construction crippled the area's economy and divided Tremé in two. It is noteworthy that, although the construction of this interstate ramp was originally an imposition on the part of city and federal governments, in succeeding years, residents appropriated the spaces underneath the bridge, making them gathering places during Second Lines and Downtown Super Sunday as well as places of quotidian socialization (Breunlin and Lewis, 2009) (Figure 6.3).

Source: Photograph by Rachel Breunlin, courtesy of the House of Dance & Feathers archive

Figure 6.3 *ReNew New Orleans second line passing under I-10 in 2006*

Bernard's presentation also addressed the future of what architects were calling the Lafitte Corridor, and the possible construction of a green commercial and residential space on the site of a major public housing facility that was closed by the HUD and the HANO directly after the storm and was slated for demolition:

> *We need different types of housing in the area, single family housing, multi family housing, elderly housing, and mixed income housing. Before HANO [Housing Authority of New Orleans] begins tearing down housing we need to make sure we have a replacement strategy… One of the things we hope to get out of this is the creation of guidelines, sustainable from an energy perspective, so that it's easy to maintain the neighbourhood, the overall neighbourhood is encouraged. Private investment is important. When we start getting the money into the city, the overall investment will happen at a faster pace. Commercial corridors can create a much better neighbourhood. Cafés, small grocery stores, drug stores, things that service the community allow for these things to pop up. As Lafitte gets reconstructed, if we think about planning holistically, as an integrated process you can get more value out of the dollar.* (Zyscovich presentation, St. Augustine Church, 13 July 2006)

At a previous presentation of this same neighbourhood recovery plan before a different audience of non-Tremé residents, Bernard Zyscovich explained the reasoning for these specific projects in the following way: 'Recovery plans need to be sold in terms of their investment potential, the federal government is much more willing to invest 5 dollars when it is going to get 25 dollars in return, than 5 dollars in mere social services' (Neighbourhoods Planning Network, 12 July 2006).

Although the Zyscovich team considered these ideas to be a reasonable approach to the neighbourhood's recovery, the overwhelming majority of residents in attendance voiced strong objections to the plan. The Zyscovich plan conceptualized neighbourhood recovery and social well-being as something to be produced through the creation of relationships between architectural structures that enhance the economic productivity of the neighbourhood. The emphasis on the creation of economic corridors and the economic revitalization of Claiborne Avenue, however, was not accompanied by stipulations that insured Tremé's displaced pre-Katrina residents would be the ones to benefit from these transformations.

The responses to this plan articulated by Tremé residents suggested they defined the neighbourhood in terms of its human landscape. People, moreover, were conceptualized by several outspoken residents as subjects who were constituted in historically specific ways through their lifelong experiences in Tremé, and were therefore irreplaceable parts of the neighbourhood. Plans that emphasized economic revitalization without the reinstitution of the neighbourhood's predominantly working class population, then, were seen as insensitive to resident perspectives on the nature of neighbourhood and social well-being. These issues came to the forefront as one resident commented: 'Armstrong Park was a plan too. One hundred and sixty four families were moved out. The plan never worked. Your proposed plan, the green space, everything sounds like the entire area is going to be commercialized. I know there are a lot of homeowners in this room. Nobody had an input as to what our neighbourhood is going to look like' (Zyscovich presentation, St. Augustine Church, 13 July 2006).

This resident referenced the construction of Louis Armstrong Park, a controversial urban renewal project carried out during the 1970s that persists in the memory of many residents as an icon of the differences between resident and planning expert visions of social well-being in New Orleans. The park's construction, for example, resulted in the permanent removal of neighbourhood residents and the construction of a park whose use by Tremé's population was restricted by gates and fences around its perimeter. While the park may be seen as an architectural improvement by some, for this resident, the permanent displacement of former neighbours still stands as an indicator of the Armstrong Park project's lack of sensitivity towards resident notions of well-being and community, and his statement suggests that he saw similar tendencies in the Zyscovich plan. Another resident followed up with the comment: 'Bring the people back that want to come back home. Open up the projects, let the people back in! Everything can't become a green space, people need housing, people need where to live!' (Zyscovich presentation, St. Augustine Church, 13 July 2006).

While many residents insisted on the importance of the reinstitution of Tremé's pre-Katrina population as the primary objective of recovery planning, the Zyscovich plan emphasized the conceptualization of the neighbourhood's urban space as a locality that is best thought of as a collection of architectonic relations that must be optimized for the investment and reproduction of capital. This is evident in statements such as '[r]ecovery plans need to be sold in terms of their investment potential', and 'if we think about planning holistically, as an integrated process, you can get more value out of the dollar'. This latter statement, in particular, is characteristic of neoliberal perspectives on urban planning, perspectives that attempt to reduce all realms of human behaviour and experience to simplistic narratives of cost–benefit analysis.

As Barrios (2009) and Ganapati and Ganapati (2009) have demonstrated, recovery plans that give precedence to cost–benefit logics over the values and culturally specific uses of space by disaster-affected populations run the risk of being seen as irrelevant, if not disruptive, by local populations. Moreover, anthropological studies of Second Line and Mardi Gras Indian parading practices in the Fourth Planning District have remarked how these ways of structuring and experiencing urban space do not operate on logics of capitalist space and time, but are, instead, mechanisms for the dissemination of community values and the production of embodied dispositions that are unique to the cultural history of New Orleans' predominantly African-American working class neighbourhoods (Lipsitz, 1988; Regis, 1999). Despite widespread opposition to the 'economic revitalization' of Tremé through the removal of I-10 and the development of North Claiborne Avenue as a major retail thoroughfare, the project remained a principal part of the City Council-Lambert plan (and, subsequently, the UNOP plan) for Tremé (UNOP, 2007).

The UNOP process

The City Council-Lambert plans were scheduled to be completed by the end of August 2006 and were intended to fulfil the participatory requirement for the city's recovery plan. In July of 2006, however, local newspapers announced a new planning process that was to supersede the City Council-Lambert plans. The process was called UNOP (Unified New Orleans Planning) and it was organized through a collaboration between the Rockefeller Foundation, the Louisiana Recovery Authority (a state agency designed to coordinate reconstruction efforts at a state level), and a local development non-profit called The Greater New Orleans Foundation (Ouroussoff, 2006).

The UNOP neighbourhood planning process was structured to take place over the course of four meetings between residents and planning teams. These meetings occurred in the autumn and winter of 2006 and early 2007. On 16 December 2006, planning team architects and residents of District 4 met at John McDonough High School on Esplanade Avenue to conduct the final discussion of the planning document. During one of the presentations by the District 4 planning teams, an architect from the St. Louis-based architecture firm, HOK, stood before a Powerpoint projection. The projection showed a graphic of the city's central neighbourhoods rendered from an aerial perspective (Figure 6.4). At the centre of the graphic was Tremé. On the image, Tremé was superimposed by translucent red arrows that crossed each other at a perpendicular angle over the neighbourhood. This image would eventually become the principal image of the District 4 final plan. As he presented the image, the architect spoke:

> *What is the potential of Louis Armstrong Park? Historically, it is very significant. Louis Armstrong is one of the most important elements of New Orleans. Armstrong must be connected to the River and to City Park. We need to consider its relationship to Jackson Square and to Iberville ... Tremé needs to be integrated into two corridors, one with the French Quarter and Lafitte, and one with Iberville.* (Unified New Orleans Planning meeting, 16 December 2006)

According to the creators of this image, neighbourhood recovery was something to be produced through the creation of spatial relationships between architectural structures that would permit the circulation of people and capital through the Fourth Planning District. But again, as in the case of the City Council-Lambert plan, some outspoken residents strongly objected to this

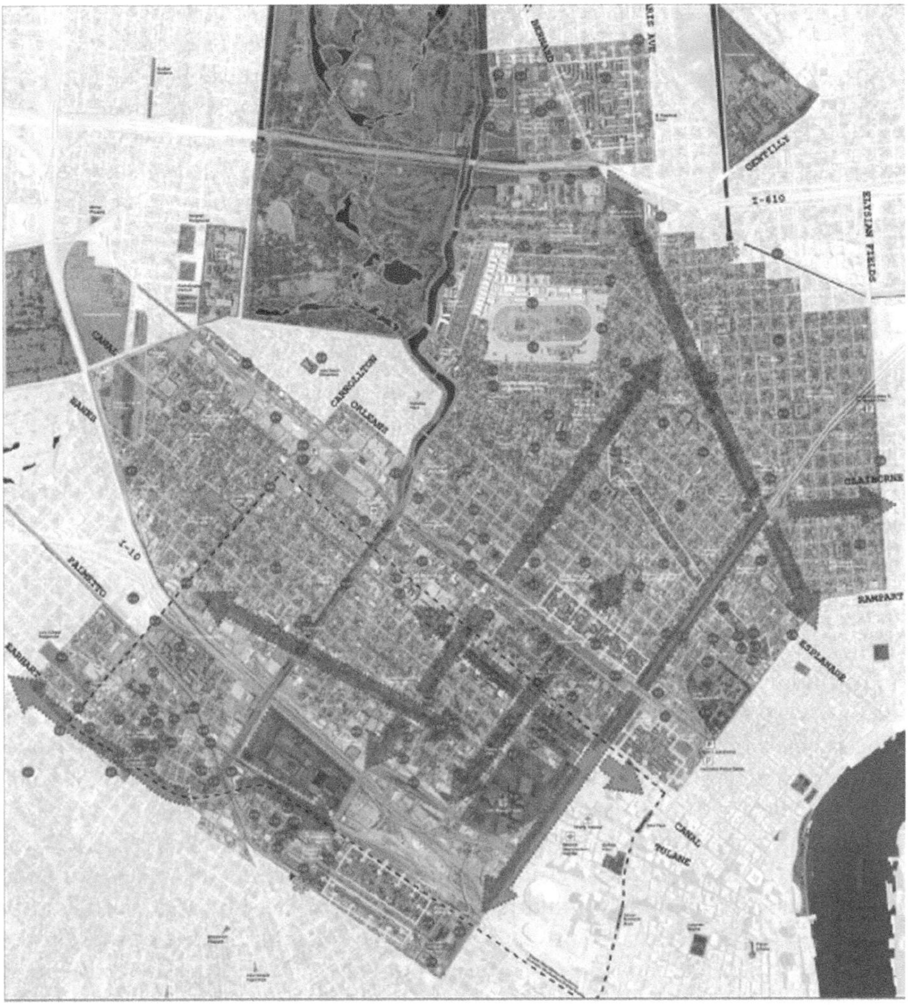

Source: UNOP (2007, p3)

Figure 6.4 *New Orleans Fourth Planning District final plan*

conceptualization of neighbourhood recovery. In response to the architect's presentation, one resident commented:

> *Two things. One is about affordable housing. Again, and again, it's been stated here in District 4, that we are very much afraid of what happened in St. Thomas [a public housing area that was unsuccessfully 'redeveloped' prior to Katrina for mixed-income housing]. Two thousand families were removed and only about 100 families returned. 'Mixed income' in the city*

> *of New Orleans, it means the same thing. There are hundreds of thousands of people waiting to come here. There have to be some things right now that say: 'How do we get people who want to come home back now.' We have discussed a number of ways of getting existing homes ready for people to come back. Somehow, something has to be put in these plans. They are renters. We need to get these people back. Where is the housing issue? The most important issue?* (Unified New Orleans Planning meeting, 16 December 2006)

This resident was concerned that, in the language of planning cities as corridors and nodes that facilitate the movement of people and capital, a discussion of what, for some residents, was the most important issue (the return of the city's displaced population and the reinstatement of the Fourth Planning District's human topography) was relegated to a secondary, if not forgotten role.

UNOP plans were completed in the spring of 2007, but the plans lacked guidelines and funding for implementation. Consequently, Mayor Ray Nagin created the Office of Recovery Management under the direction of Dr Edward Blakeley. The purpose of this new office, as explained by a staff member during an ethnographic interview was to 'plan to implement', not create, new plans. Officially, the creation of the Office of Recovery Management was not meant to invalidate the UNOP plans, but planning documents produced by the Office of Recovery Management have differed significantly from UNOP documents. The former, for example, have relied on the designation of 17 Target Zones in the city, where imminent domain laws may be used to claim urban space that will be used as concentrated nodes of economic investment. The logic of the 17 Target Zones was that, rather than spending public reconstruction funds for the immediate provision of public services needed by New Orleans' most disaster-vulnerable populations (public hospitals, housing, schools), the people of New Orleans would be better served through the use of these funds to create incentives for out-of-state investment. Out-of-state investment, in turn, would create the tax revenue that, in a hypothetical future, could then be invested in these public services. In December of 2008, however, the Office of Recovery Management had yet to find suitable investors for such ventures (Donze, 2008). Consequently, the city of New Orleans stood to lose up to one billion US dollars in federal reconstruction funds, and hurricane-devastated neighbourhoods like the lower 9th ward remained critically underserved in terms of public services and infrastructure (Hill, 2008).

Discussion

The ethnography of disaster recovery planning in New Orleans' Fourth District reveals that there were critical differences in the ways some residents and planning experts conceptualized the nature of neighbourhoods and the practices necessary to create a sense of recovery. While many residents insisted on the conceptualization of neighbourhood recovery as the reinstatement of the city's disaster-displaced working class population that is responsible for the quotidian and ritual practices that shape New Orleanians' embodied dispositions, architects and planners conceptualized neighbourhoods as a collection of architectonic structures and relationships optimized for capital reproduction. Furthermore, rather than adhering to official representations of disaster recovery planning as a process where professional planners simply format and disseminate resident ideas about neighbourhood recovery, planning experts introduced novel ways of conceptualizing neighbourhood recovery and actively resisted requests on the part of many residents for alternative means of thinking about and experiencing urban space and recovery.

The emphasis of planning experts on the conceptualization of neighbourhood recovery as something to be produced through capital investment in New Orleans' Fourth Planning District must be problematized in a number of ways. First, the idea of 'neighbourhood as a mechanism of capital reproduction' was presented by planning experts as naturalized socio-economic matter-of-fact that was not open to negotiation. The anthropological literature, however, has demonstrated that capital is not a universal logic that underlies the practices of all societies. Instead, it is a culturally specific way of thinking about and engaging the social and material world. Second, because it is a cultural convention, neoliberal thought is always interpreted by human actors on the basis of their own cultural frameworks, subverting the possibility of the manifestation of a global capitalism that operates in a standardized fashion. Finally, urban planning policies that rely on neoliberal tenets of 'urban space as a mechanism of capital reproduction' and the utilization of public resources for the encouragement of private investment run the risk of increasing the kinds of social inequities that manifest so dramatically during disasters. In the aftermath of Hurricane Katrina, the city of New Orleans has witnessed a dramatic surge in property values and rental rates, the demolition of four major public housing developments, a re-emerging homeless population, and a crisis among non-profit organizations that have borne the burden of providing assistance to disaster-affected populations in obtaining basic services (housing, utilities, health-care) (Olson, 2007). As Allen and Maret (Chapter 7) show, neoliberal logics of disaster recovery have also delayed the provision of critical social services like public schooling in Tremé, enhancing socio-economic differences between those who can afford private education, and those who cannot. The conclusions of this study are best summarized in two recommendations for recovery planning practitioners:

1 In order for participatory planning to be truly participatory, planning experts must recognize the cultural, political and historical particularity of the epistemological assumptions implicit in the ways they conceptualize social well-being, recovery and the nature of urban space.
2 Expert knowledge must be open for negotiation with participant views of neighbourhood recovery and social well-being. The rigid application of expert knowledge has, time and time again, proven to threaten environmental sustainability and to create conditions of social marginalization.

References

Bankoff, G. and Hilhorst, D. (2004) 'Introduction: Mapping Vulnerability', in G. Bankoff and D. Hilhorst (eds) *Mapping Vulnerability*, Earthscan, London

Barrios, R. E. (2009) 'Tin Roofs, Cinder Blocks, and the Salvatrucha Gang: The Semiotic Material Production of Crisis in Post-Hurricane Mitch Reconstruction', in M. O. Ensor (ed) *The Legacy of Mitch: Lessons from Post-Disaster Reconstruction in Honduras*, University of Arizona Press, Tucson, AZ

Blaikie, P., Cannon, T., Davis, I. and Wisner, B. (1994) *At Risk: Natural Hazards, People's Vulnerability, and Disasters*, Routledge, New York

Bourdieu, P. (1977) *Outline of a Theory of Practice*, Cambridge University Press, Cambridge

Breunlin, R. and Lewis, R. W. (2009) *The House of Dance and Feathers*, University of New Orleans Press, New Orleans, LA

Breunlin, R. and Regis, H. (2006) 'Putting the ninth ward on the map: Race, place, and transformation in Desire, New Orleans', *American Anthropologist*, vol 108, no 4, pp744–764

Caldeira, T. and Holston, J. (2005) 'State and Urban Space in Brazil: From Modernist Planning to Democratic Interventions', in A. Ong and S. J. Collier (eds) *Global Assemblages: Technology, Politics, and Ethics as Anthropological Problems*, Blackwell Publishing, Malden, MA

Campanella, R. (2006) *Geographies of New Orleans: Urban Fabrics Before the Storm*, Centre for Louisiana Studies, Lafayette, LA

CUPA (1995) *Enhancing the Sense of Place in Tremé: Mechanisms for Preserving a Unique, Historic Neighbourhood*, University of New Orleans College of Urban and Public Affairs, New Orleans, LA

Davis, M. (2006) 'New Orleans rebuilding unveiled', news.bbc.co.uk/2/hr/americas/4599832.stm, accessed 15 January 2007

di Leonardo, M. (2008) 'Introduction: New global and American landscapes of inequality', in J. Collins, M. di Leonardo, and B. Williams (eds) *New Landscapes of Inequality: Neoliberalism and the Erosion of Democracy in America*, SAR Press, Santa Fe, NM, pp3–20

Donze, F. (2008) 'N.O. recovery czar outlines budget', www.nola.com/news/t-p/index.ssf?/base/news-9/1227334995282870.xml&coll=1, accessed 22 November 2008

Douglas, M. (1992) *Risk and Blame: Essays in Cultural Theory*, Routledge, New York

Dowdy, S. L. (2008) *Building the Devil's Empire: French Colonial New Orleans*, University of Chicago Press, Chicago, IL

Escobar, A. (1995) *Encountering Development*, Princeton University Press, Princeton, NJ

Ganapati, N. E. and Ganapati, S. (2009) 'Enabling participatory planning after disasters: A case study of the World Bank's housing reconstruction in Turkey', *Journal of the American Planning Association*, vol 75, no 1, pp41–59

Gerber, C. (2006) 'In uncertain New Orleans, failure of leadership', *USA Today*, 26 August, p10A

Holston, J. (1989) *The Modernist City: An Anthropological Critique of Brasilia*, University of Chicago Press, Chicago, IL

Hill, L. (2008) 'Interview of Lance Hill on Hurricane Katrina and racial and class equity issues', communitygumbo.blogspot.com/2008/01/1262008-community-gumbo.html, accessed 20 October 2009

Irazábal, C. and Neville, J. (2007) 'Neighbourhoods in the lead: Grassroots planning for social transformation in post-Katrina New Orleans?', *Planning, Practice & Research*, vol 22, no 2, pp131–153

Johnson, J. (1992) 'Colonial New Orleans: A Fragment of the Eighteenth-Century French Ethos', in A. Hirsch and J. Logsdon (eds) *Creole New Orleans: Race and Americanization*, Louisiana State University Press, Baton Rouge, LA

Lacho, K. J. and Fox, K. (2001) *An Analysis of an Inner-city Neighbourhood: Tremé – Past, Present, and Future*, University of Central Arkansas Small Business Advancement National Centre, Conway, AR

Lipsitz, G. (1988) 'Mardi Gras Indians: Carnival and counter-narrative in Black New Orleans' *Cultural Critique*, vol 10, pp99–121

Lipsitz, G. (2006) 'Learning from New Orleans: The social warrant of hostile privatism and competitive consumer citizenship', *Cultural Anthropology*, vol 21, no 3, pp451–468

Mitchell, T. (2002) *Rule of Experts: Egypt, Techno-Politics, Modernity*, University of California Press, Berkeley, CA

New Orleans Renovation (2006) 'Historic Faubourg Tremé Association takes aim at blight, crime and grime', neworleansrenovation.blogspot.com/2006/09/historic-faubourg-tremeassociation.html, accessed 3 July 2007

Nossiter, A. (2007) 'Harry Lee, outspoken Louisiana sheriff, dies at 75', www.nytimes.com/2007/10/02/us/02lee.html?sq=harry%20lee% 20jefferson%20parish&st=nyt&adxnnl=1&scp=5&adxnnlx=1215108388- 9xKRWce+qRY5XlImTZjumA, accessed 8 February 2008

Oliver-Smith, A. (1999) 'What is a Disaster? Anthropological Perspectives on a Persistent Question', in A. Oliver-Smith and S. Hoffman (eds) *The Angry Earth: Disaster in Anthropological Perspective*, Routledge, New York

Olson, L. (2007) *Performance Evaluation and Recommendations for The Greater New Orleans Disaster Recovery Partnership and the Long-term Recovery Committees of Plaquemines Parish, Jefferson Parish, Orleans Parish, St. Bernard Parish*, The George Washington University Institute of Crisis, Disaster, and Risk Management, Washington DC

Ong, A. (2006) *Neoliberalism as Exception: Mutations in Citizenship and Sovereignty*, Duke University Press, Durham, NC

Ouroussoff, N. (2006) 'In New Orleans, each resident is master of plan to rebuild', *The New York Times*, 8 August, section E, p1

Rabinow, P. (1995) *French Modern: Forms and Norms of the Social Environment*, University of Chicago Press, Chicago, IL

Regis, H. (1999) 'Second lines, minstrelsy, and the contested landscapes of New Orleans Afro-Creole festivals', *Cultural Anthropology*, vol 14, no 4, pp472–504

Scott, J. C. (1999) *Seeing Like a State: How Certain Schemes to Improve the Human Condition Have Failed*, Yale University Press, New Haven, CT

Schuller, M. (2008) 'Deconstructing the disaster after the disaster: Conceptualizing disaster capitalism' in N. Gunewardena and M. Schuller (eds) *Capitalizing on Catastrophe: Neoliberal Strategies in Disaster Reconstruction*, Alta Mira Press, Lanham, MD, pp17–27

Shaw, R., Gupta, M. and Sarma, A. (2003) 'Community recovery and its sustainability: Lessons from Gujarat earthquake of India', *Australian Journal of Emergency Management*, vol 18, no 2, pp28–34

Sorant, P., Whelan, R. and Young, A. (1984) 'City profile: New Orleans', *Cities*, 1 May

Tanaka, T., Abramson, D. L. and Yamazaki, Y. (2009) 'Using GIS in community design charrettes: Lessons from a Japan–U.S. collaboration in earthquake recovery and mitigation planning for Kobe', *Habitat International*, vol 33, pp310–318

UNOP (2007) *District 4 Recovery Plan: Final District Neighbourhood Plan*, Unified New Orleans Planning, New Orleans, LA

US Census Bureau (2000) *US Decennial Census*, factfinder.census.gov/, accessed 17 May 2008

Warner, C. (2006) 'Council wants its planners retained', www.nola.com/news/t-p/frontpage/index.ssf?/base/news-6/1151734586140970,xml&coll=1, accessed 23 August 2007

Warner, C. (2007) 'Getting together', www.nola.com/news/t-p/frontpage/index.ssf?/base/news-7/117238740432060.xml&coll=1&thispage=1, accessed 25 February 2007

Wisner, B. (1995) 'Bridging "expert" and "local" knowledge for counter-disaster planning in urban South Africa', *GeoJournal*, vol 37, no 3, pp335–348

7
Rebuilding the Historic Tremé Neighbourhood: Lessons in the Repatriation of New Orleans

Barbara L. Allen and Isabelle Thomas Maret

Hurricane Katrina quickly became iconic among environmental and social justice scholars and activists for its revelatory images and immediate uncovering of racial disparities in the US. The storm, the flood and the incompetent response by government agencies have become a part of our collective memory. By far the largest recipients of these failures have been the poor and minority residents of New Orleans driven out of their homes in record numbers. The initial US government failures of the Federal Emergency Management Agency (FEMA), followed by similarly poorly organized programmes, such as the Road Home programme, designed to compensate homeowners in rebuilding, left New Orleanians looking elsewhere for help. Fortunately, other programmes, under the auspices of non-governmental organizations (NGOs), religious groups and even, in this example, a competent, small government agency began to fill in where FEMA and large programmes had failed. While this amalgam of diverse groups has not been a replacement for a well-run, large-scale rebuilding programme citywide, it has provided much-needed help to more needy neighbourhoods. In order to understand the difficulties as well as the successes and resilience of New Orleans' residents and the synergies of outside aid groups, we chose to study one important historic neighbourhood in the city – Tremé.

Before beginning the discussion, it is important to place our case study within a set of theories to better understand and analyse the dynamics of the stresses and successes of rebuilding in a social context. The interdisciplinary field of Science and Technology Studies (STS) provides two theoretical tools that, reinterpreted, are particularly salient in our research: the public understanding of science (PUS) and the new political sociology of science (NPSS).

PUS is a broad and multifaceted approach to understanding the democratic tendencies in scientific understanding and policy application (Irwin and Wynne, 1996). One focus of the early work was on the necessary hybridity of science and what counts as policy-relevant knowledge in pluralistic societies (Irwin and Michael, 2003). In the environmental arena, where contention is rampant, explicit permeability of traditional science is required to move forward policy-making agendas (Irwin, 1995; Fischer, 2000). While PUS has more typically been used as an analytical lens into the discourse and practices of science, we suggest in our research that it is usefully re-visioned as a tool for understanding the public understanding, engagement and use of technologies as well. The post-disaster rebuilding of New Orleans involved an array of technologies, from those of relief and assistance management to actual techniques of home renovation.

The second analytical lens, NPSS, addresses the insularity of case studies, in general. It seeks to connect case studies to the broader cultural, political and institutional dynamics that have led to differential availability of, and access to, power domains in knowledge production. Often scholars working in this emerging field examine how laws, rules, codes, organizations and networks impact power and knowledge. It makes visible the politics of institutional arrangements as well as positive (and negative) policy fulcrums for change. By drawing attention to the aggregative and distributive processes in knowledge-making it enables the building of comparative case studies, ones that can build broader hypotheses and explanations (Frickel and Moore, 2006). Extrapolation of this 'science' lens to the technological arena of post-disaster rebuilding can better illuminate both inequities and positive synergies and outcomes.

The Tremé neighbourhood – A focal point of African-American and Creole culture

Tremé is one of the oldest African-American neighbourhoods in the US, dating back to the late 18th century when free people of colour and refugees from what is now Haiti and Santo Domingo populated the area (Figure 7.1).

Annexed by the city in 1812, it soon emerged as one of the most prosperous and politically active black communities in the US. During the 19th century blacks, whites and mixed-race people lived side by side creating a Creole culture, the very culture that defines New Orleans 'exceptionality' today. One of many notable residents was Homer Plessy who became the plaintiff in the landmark US Supreme Court case, Plessy v. Ferguson, which reinforced and institutionalized segregation in their 1896 ruling. Tremé is also the birthplace of Louis Armstrong, Jelly Roll Morton and other jazz greats.

Within its bounds is Congo Square, a gathering place and market where free blacks and slaves used to congregate in the antebellum south and probably the birthplace of jazz. Additionally, Tremé is known for performative cultural events that mark space in distinctive and sometimes subversive ways that was to become a focus of conflict post-Katrina (Johnson, 1991; Allen,

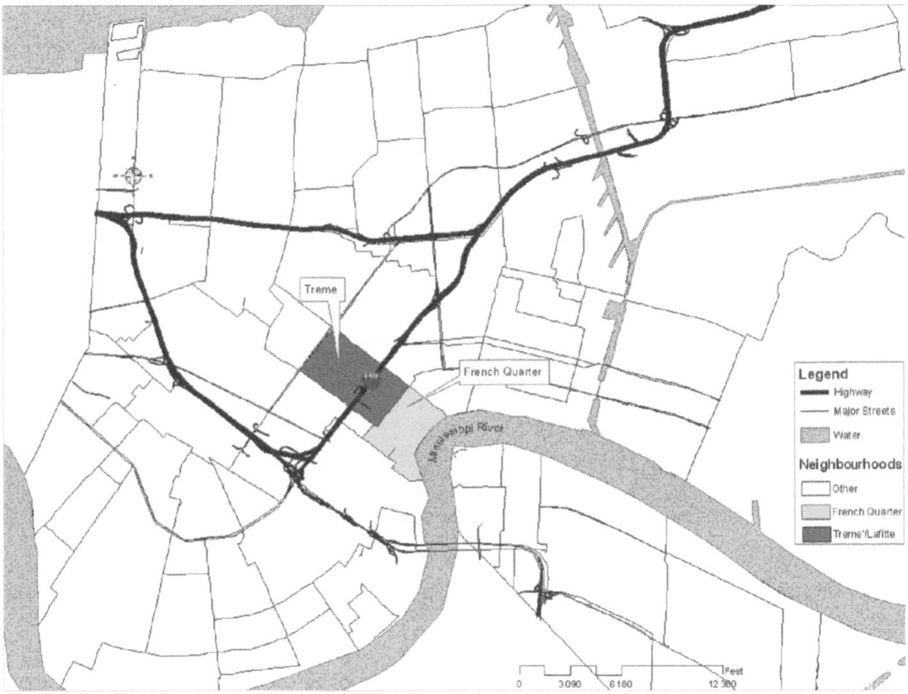

Source: I. Maret

Figure 7.1 *Map of Tremé*

2007a). There are also a number of historically significant museums, churches, parks and other 19th-century structures, marking it as both an incubator and centre of African-American culture in the city (Figure 7.2).

In the 1960s a series of urban design and planning assaults on the community led to its decline. Tremé was divided by an interstate highway, I-10 (Figure 7.3), which effectively erased its business corridor along Claiborne Avenue.[1] Another unfortunate project was the building of the Municipal Auditorium and Louis Armstrong Park which is now a walled off part of the desirable French Quarter side of Tremé in an underutilized 'civic development' (Figure 7.4). While urban renewal created urban blight and economic hardship in the neighbourhood, it also proved a catalyst for strengthening community groups dedicated to preserving Tremé and its traditions. The community realized that Tremé's identity was both its architectural history (Figure 7.5) as well as its cultural-spatial practices (Crutcher, 2001) (Figure 7.6).

Prior to Katrina, many of the homes and buildings in Tremé were in mild to moderate disrepair, due to a lack of community resources. Because of this, the neighbourhood sustained significant damage from wind and rain (Figure 7.7) during and after the storm though it received only a small amount

Source: Photo by I. Maret

Figure 7.2 *New Orleans African-American Museum of Art in the Meilleur-Goldthwaite House, c. 1828–29*

Source: Photo by I. Maret

Figure 7.3 *Interstate I-10 cutting Tremé's community*

Source: Photo by I. Maret

Figure 7.4 *Louis Armstrong Park, a walled enclave in the Tremé neighbourhood*

Source: Photo by I. Maret

Figure 7.5 *A typical Creole cottage (c. 1850s) in Tremé being repaired by the Qatar Tremé Renewal Project in collaboration with the owner, Mr Thomas*

Source: Photo by I. Maret

Figure 7.6 *Example of elaborate Mardi Gras Indian costumes from the Backstreet Museum*

of damage from flooding.[2] An estimated 52 per cent of the homes received 'severe damage' (over $30,000) and an additional 17 per cent received 'major damage' (between $5200 and $30,000) (Greater New Orleans Community

Source: Photo by I. Maret

Figure 7.7 *House with significant damage being repaired by the Qatar Tremé Renewal Project*

Data Center, 2010a). Placing it in further peril, Tremé is located on high ground adjacent to the French Quarter and is currently a prime target for both new development and gentrification (Figure 7.8).

Source: Photo by I. Maret

Figure 7.8 *New condominiums on Rampart Street, which divides Tremé from the French Quarter*

Repatriation complications 1: The racial politics of music in the streets

Prior to Katrina, the Lafitte Housing Project in Tremé housed over 800 children, almost half of the projects' total population. During the day hundreds more were in residence, being cared for by relatives and grandparents (the informal daycare economy) while their parents worked (Reckdahl, 2006b). The place was filled with kids marching and playing anything they could get their hands on as a drum. According to Jerome Smith, New Orleans' musical traditions begin with children playing together and with their older peers.[3] Smith runs *Tambourine and Fan* out of the Tremé Community Centre. It is an organization that he founded decades ago dedicated to bringing the music and culture of New Orleans to children through summer camps and after-school programmes. Before Katrina his summer camps used to serve between 500 and 600 children, but 2 years after, only 150 students attended. Besides the lack of children, Smith points to another issue affecting the music traditions in New Orleans. There are comparatively so few schools open in the city, that students are bussed to and from school each day making programmes like his difficult to continue. 'Before Katrina', says Smith, 'kids went to the same elementary, junior high and high school. After school whole blocks used to come alive with music – kids playing tubes, sticks, cans and boxes. Older kids wearing band uniforms would be cheered on by the neighbourhood on their way to practice. They were a source of pride for the community.'[4]

There are other signs that local music traditions are under duress. In November 2007 the New Orleans Recovery School District closed Craig

Elementary School in Tremé, a school that has been considered a focal point in this community for generations (Simon, 2007). Moisture and health problems were cited as reasons for its closing. Students were bussed to New Orleans East to modular buildings, impacting both after-school music programmes such as Tambourine and Fan as well as the ability of kids to form spontaneous music groups as part of neighbourhood play. The school reopened in early 2010 amidst a contentious debate about who would run it, the city pushing hard for a specific non-local private charter firm (Carr, 2010). Questions about music and other cultural traditions being taught and passed on to future generations remain.

Tremé is probably most famously known as the home of the jazz funeral and the second line tradition, Creole and African-American customs dating back to at least the 19th century. The jazz funeral is a music procession through the neighbourhood after the church funeral. This involves musicians and family members following the deceased in a horse-drawn hearse, parading down the street to the cemetery or for about ten blocks if the cemetery is far away. The music is typically sombre at first and then picks up tempo after the burial or when the hearse leaves the parade, as the people let their beloved go. Along the way, neighbours, friends and onlookers often join in the celebration of the person's life forming what is called a 'second line'. As they follow along they sometimes twirl a parasol or wave handkerchiefs, dancing in a style referred to as 'second lining'. These parades are a common sight in Tremé and are an important part of the culture of this neighbourhood (Spitzer, 2006).

Because of gentrification, hastened by Katrina, the demographics of the neighbourhood are changing, causing some tension between cultural groups.[5] On Monday, 1 October 2007, a second-line memorial procession in Tremé, celebrating the life of musician Kerwin James who died of complications from a stroke after Katrina, was brought to an abrupt end by police. Twenty police cars responded to noise complaints from some of the neighbours, surrounding the mourners and hauling some of the musicians away in handcuffs. Two musicians were charged with misdemeanours, specifically 'disturbing the peace and parading without a permit'. Tuesday, the following night, the mourners paraded again, this time without incident, escorted by the New Orleans Police Department as they had obtained a permit that day.

The over-reaction by police to a cultural practice that had previously been allowed to happen without interference (or permits) was a source of outrage in the Tremé community. One community member, Beverly Curry explained that for over a century there have been second-line parades for musicians 'from the day they pass until the day they're put in the ground' (Reckdahl, 2007). Suddenly enforcing a never-enforced permit statute on impromptu second-line parades, seemed to some to be an attempt to control or even limit these sorts of traditional cultural activities. Louis Charbonnet, the fourth-generation owner and director of the main funeral home in Tremé agreed it was a cop-out, 'but sometimes you have to do what you have to do'(Reckdahl, 2007).

Describing other complicating factors, Charbonnet explained that since the storm there have been many more funerals, some for people whom he considers storm victims. There are those that die from stress or simply give up, and those that quit taking their medication, or lose access to their doctors and medication. There were heart attacks and depression and a host of respiratory illnesses, all complicating someone with fragile health.[6] Charbonnet's anecdotal observations match the informal data that Kevin Stephens, the Director of the New Orleans Heath Department had been collecting using the obituary section of the local newspaper in order to ascertain deaths that occurred out of the city and state as well as locally. His study noted a 47 per cent increase in the mortality rate in New Orleans in the six months following Katrina (American Medical Association, 2008). Additionally, many of the African-American funeral homes have been closed since the storm, including his brother's establishment in the lower 9th ward, making Charbonnet's home one of the few operating in the city. Business has been brisk and the number of jazz funerals and second-line parades, plenty.

According to Charbonnet, the unwritten law was that a jazz funeral in Tremé was a ten-block parade lasting about 30 minutes with the deceased pulled in a horse-drawn hearse to one of the nearby cemeteries. When musicians died, there was often a short informal gathering or spontaneous grieving every night until the funeral. However, many long-time residents agree that the new police presence and permit enforcement is due to the new white residents voicing noise and safety concerns. According to resident Curry, 'They want to live in Tremé, but they want it for their ways of living. I tell them when someone dies you're going to hear a band. When they say they are concerned about crowds and noise I say, "You found us doing this – this is our way"'(Reckdahl, 2007). Jerome Smith put it a bit more bluntly, framing the issue in the performative nature of local culture: 'What happened two weeks ago was because it [the jazz funeral] was not white. Neighbourhoods are the spirit of New Orleans – they are not just a collection of houses – and music is not only in those that carry instruments!'[7]

Repatriation complications 2: Mired in controversy – the Lafitte Public Housing Project

Long before Katrina hit, the Housing Authority of New Orleans (HANO), like other public housing administrations around the country, had been in the process of dismantling and re-engineering housing for the poor. Under a federal programme, called HOPE VI, bunker-style buildings housing dense concentrations of poor people, would be torn down and mixed-income housing would be built that was visually integrated into the surrounding residential community. Ideally the displaced poor would then be welcomed back to a better, safer community instead of being warehoused in 'the projects'. This was the ideal – the reality had a few holes in it.

The first problem was where to house people while the new housing was being built. Theoretically, they would be given vouchers enabling them to move into market rental housing. However, not all landlords were keen on renting to the poor, thus proving more difficult in actuality. Second, the mixed-income housing that was eventually built was never designed to house the same number of poor people, assuming that some would remain in private rentals, having the social consequence of lowering the density of poverty in any one neighbourhood.

After Katrina, only a portion of one of the smaller housing projects, Iberville, next to the French Quarter, was reopened. Over two years after the storm, the remaining 'Big Four' containing 4500 units (B.W. Cooper, C.J. Peete, St. Bernard and Lafitte) were slated for demolition, backed by a vote from the City Council. However, the mayor refused to sign the demolition permit for one of them, the Lafitte Housing Project, until the Department of Housing and Urban Development (HUD) produced a firm proposal and financing for how it would be redeveloped. The Lafitte project was located in Tremé, at the foot of the natural levee, not below sea level and took on very little water during the flood – to the naked eye it appeared undamaged.

By this time there was additional pressure brought by the media, civil rights groups and the displaced poor themselves to provide housing for them to move back to New Orleans as well. Why not quickly clean up the 896 units at Lafitte and allow the approximately 2000 residents to return? After all, internal HUD and HANO memos revealed that modernizing the project would cost less than demolition and rebuilding (Reckdahl, 2006a).[8] Rather than destroy what some termed historic structures, built in the style of the Pontalba Apartment buildings on Jackson Square, making more landfill debris, organizations such as the National Trust for Historic Preservation suggested these building be retained. According to Richard Moe, president of the National Trust,

> *...it is possible and desirable to use the historic housing projects as the basis for carefully planned mixed-income redevelopment in which new construction is blended with rehabilitated older buildings... However, our proposals have been drowned out by demands that almost every building be eradicated – as if destroying the structures themselves would eliminate the failed social and housing policies they symbolize.* (Foster, 2008)

But the mayor, HANO and HUD had other ideas. The state and federal government awarded Providence Community Housing (a post-Katrina initiative of Catholic Charities) in collaboration with Enterprise Community Partners (an affordable housing non-profit) close to $200 million in funds and tax credits to redevelop the property and the surrounding area. Their idea was to demolish all but a few of the Lafitte structures and then use the remaining property for street-oriented townhouse style housing that would be integrated with neighbourhood. Additionally, they had purchased over 200 vacant lots from the city in this area in which they planned in-fill housing, such as shotguns, cottages and small apartments (Providence Community Housing, n.d.). The Providence-Enterprise group proposed to bring new life and housing to the poor, low and lower-middle income area of Tremé, that was located in a sort of no man's land north of I-10.

The non-profit housing group has made clear that former residents will be given first right of return. According to a survey done by All Congregations Together of 400 of the former residents of Lafitte, over 90 per cent want to return. (Reckdahl, 2006a) There is still much scepticism and concern about the time it will take to demolish and rebuild the Lafitte site as a mixed-income project and if everyone that wants to will be able to return to the Lafitte neighbourhood of Tremé. The longer people are away, the less likely it is that they will move back.

The difficult road home

In an effort to distribute the $10 billion that the federal government allocated to assist Louisiana property owners, the state contracted with ICF International, a company based in Fairfax, Virginia, to help them decide the best way to disburse the funds. The idea was that ICF was to develop a programme that would later be put out for competitive bid in order to select the best programme administration firm. While developing a blueprint for what was to become the state's now infamous Road Home Program (RHP), and realizing the lucrative nature of the programmes' administration contract, ICF decided to bid for the contract itself. Realizing their decision may give the impression of having an unfair advantage, ICF ended their earlier contract (worth $900,000) but not until the RHP was mostly developed. They were eventually awarded over $900 million dollars to administer the $10 billion in federal funding, on one of the largest non-construction contracts in state history (Sturgis, 2006).

From the beginning there were questions regarding the decisions made by the RHP in consultation with the governor, Kathleen Blanco, and other high-ranking state officials. Most notable was the dual decision to have programme incentives that favoured returning and rebuilding while at the same time closely guarding disbursements from the programme in a way that appeared paternalistic to many residents. The RHP provided larger grants to those that remained and rebuilt thus penalizing those that chose to move. Secondly, rather than releasing lump sum payment checks of up to $150,000 to homeowners to use in any way they wished (as in Mississippi), the Louisiana Recovery Authority who oversees the RHP decided that homeowner disbursement accounts would be set up in lieu of checks. Mortgage lenders would monitor these accounts and closing agents would police the spending, making sure the funds went toward rebuilding costs. This decision made Louisiana's plan more costly and complex and rebuilding would take considerably longer than the damaged coastal towns of Mississippi (Lovett, 2006).

Over time it became clear that the extra steps the RHP made to ensure accountability and encourage repatriation was having the effect of dramatically slowing down the rebuilding effort and engendering a great deal of frustration among residents already fatigued from the tragedy of their losses. Pressure from residents eventually led to no-strings lump sum payments being awarded to homeowners without mortgages, which made up about 30 per cent of the pool of applicants. Additionally, because the programme was more of a 'rebuilding' programme than a 'compensation' programme (like in Mississippi) it could be subject to environmental reviews, fair housing and wage laws, further complicating the already Byzantine programme. Anger over the slow and cumbersome RHP was also a contributing factor in Governor Blanco's decision not to seek re-election (Warner, 2007).

Citizens rallied and formed many support groups and created websites to help RHP victims, one of the most active being the 'New Orleans Wiki Citizens' Road Home Action Team' (CHAT). They lobbied, sometimes successfully, for changes and offered advice on how to navigate the ICF-led appeals process. ICF forcefully directed citizens to settle their problems on the phone with company representatives and made it very difficult for homeowners to file formal appeals. For example, a year and a half after the flood there were about 11,000 unresolved informal complaints and only 5 had formally appealed. The informal appeals process had no timelines for dispute resolution whereas a formal complaint had to be resolved within 30 days (Hammer, 2007). In mid-2008 the RAND Corporation published a revealing study of the RHP titled, 'Timely Assistance: Evaluating the Speed of Road Home Grantmaking'. In their research they discovered the mean grant wait time was 251 days and many waited for over a year (Eden and Boren, 2008). Homeowners with condos or mobile homes, or those that had some form of insurance waited even longer. As of June 2008, almost three years after the storm and two years after the RHP began, about one third of the more than 150,000 eligible applicants were still waiting for funds (Road Home Program, 2008).

Although the total allowable amount of each rebuilding grant was \$150,000 or the extent of the damage, whichever was lower, the average award disbursed as of June 2008 was about \$59,000. A number of other factors decreased the amount of money awarded to homeowners as well as increased the time it took for them to receive the money. If the homeowner did not have insurance, the grant was reduced 30 per cent and if they moved out of state after the flood and wanted to sell their home, the grant was further reduced. Any rebuilding costs provided by the homeowners' insurance or any other charities or grant programmes was also deducted from the RHP grant. One contentious issue has been the way in which the RHP calculates the grants: it was either calculated based on a home's pre-storm value or the cost to rebuild, whichever is less. This has led to much lower payments in poor and working class neighbourhoods even though the cost per square foot to rebuild was similar across much of New Orleans existing housing stock (Hammer, 2008).[9] For low-income homeowners there was an additional \$50,000 grant that homeowners could apply for from the RHP. They could also augment their grants with loans from the Small Business Administration (SBA) and banks. Yet another confusing and unfortunate glitch in the system was that the RHP's grants were not taxable unless the homeowner reported their home damage as a casualty loss on their 2005 income taxes, a loss that would have been reported before citizens knew the details of the RHP, which began taking claims in the summer of 2006. Many people did report their losses to the IRS in 2005, thus making their RHP compensation taxable.

Success in Synergy 1: Volunteer help – Rebuilding Together

In the months following the storm many volunteers flooded into the city to offer help. These came in a variety of shapes and sizes from church groups to Boy Scout troops as the nation looked on the plight of New Orleanians with anger and dismay wanting to offer some kind of help. In the beginning the plethora of help groups were poorly coordinated and sometimes put volunteers at risk in their enthusiasm to get the rebuilding process started without taking proper precautions. Other times groups, while having good intentions, actually did more damage to historic homes, unaware of what should and could be saved when gutting older housing stock (Allen, 2007b,c).

Prior to Katrina, the Preservation Resource Center (PRC) in New Orleans had run a local volunteer programme they called 'Christmas in April/October' repairing the homes of low-income elderly and disabled residents. After the storm, the programme changed its name to Rebuilding Together (RT) and using non-local volunteers they began renovating flood-damaged homes of the same population of homeowners (Rebuilding Together New Orleans 2010).[10] In order to conserve resources, the organization targeted five low-income and working class historic neighbourhoods. Like Habitat for Humanity, RT serves as a distribution centre for volunteers from all over the US and they carefully matched elderly and disabled homeowners in need with groups of volunteers such that both parties had a positive experience in the rebuilding process.

The five neighbourhoods they initially targeted for storm renovation approached RT for help, a necessary element for involvement of the organization. RT had a focus on ageing in place, and given the extended family nature of many New Orleanians' living situations, the inclusion of elderly family members often facilitated both childcare and elder care in an affordable fashion. Many older vernacular houses in New Orleans are doubles (two units) often bringing multiple generations together. Once they received rebuilding requests, they mapped out all the eligible projects in an attempt to base the final acceptance of projects on locational synergy to best allocate resources. Additionally they consulted with community groups in the neighbourhoods as they preferred to target key people and leaders in communities for repatriation help in order to foster the people-infrastructure that would encourage others to rebuild as well.

As of June 2008, RT had completed 70 houses using almost 5500 volunteers coordinated by staff and construction managers paid by outside agencies such as Vista and AmeriCorps. They had one full-time employee in charge of volunteers and another one that acted as a construction manager. They managed up to 100 volunteers a day and scheduled three build weeks a month with a one-week down time to allow for organizational assessments and supplies.[11] While RT prioritized quality of life as the main criteria for construction design decisions, being part of the PRC, they were also interested in historic preservation of the city's many older neighbourhoods. Thus they incorporated selective salvage into their rebuilding plans. For example, they tore down 125 historic houses that were beyond repair and the materials were warehoused and used when rehabilitating other older homes. In only a few years the organization expanded and by 2010 they had 27 full-time employees, 41 AmeriCorps members and planned to rebuild 160 homes using 6000 volunteers (Rebuilding Together New Orleans, 2010).

RT also shared its list of projects with other NGOs and agencies doing rebuilding work in New Orleans. Some NGOs, for example, did not do finish work, so sometimes projects were referred to RT who came in and did cabinetry and painting. Likewise they have allied themselves with groups such as the State Historic Preservation Office (SHPO) and Providence Housing (part of Catholic Charities) and together they developed a database so that more synergistic rebuilding assistance can take place. Like other NGOs in the city, they packaged their monetary assistance as a soft-second 'forgivable' mortgage so that the project value was not subtracted from citizens' RHP money, nor were the homeowners taxed on the money.

More Synergies 1: An exemplary government rebuilding programme

New Orleans had a vibrant and powerful historic preservation community and had the highest number of historic structures per person of any city in the US. Prior to the storm, the Historic District Landmark Commission had over 15 employees and oversaw the 20 historic districts in the city. Additionally, the French Quarter Commission had another 12 members, which was specifically tasked with the oversight of building and reconstruction in the French Quarter. After the storm, these committees and their staff were immediately laid off save for a handful of people in the New Orleans planning department, raising fears with preservationists that developers would be able to build in a historically insensitive fashion throughout the city. While the jury is still out on the battles between developers and preservationists in the post-Katrina era, what is clear is that the strength of the preservation movement in the city reaped some noticeable rewards.

In October 2006, a little over a year after Katrina, the US Congress awarded 18.5 million dollars to the Louisiana State Historic Preservation Office (SHPO) to help rebuild damaged historic homes and neighbourhoods across south Louisiana. The project was funded through the National Park Service (NPS), a division of the US Department of the Interior who aided the state in establishing both historic and rebuilding guidelines. The deadline in December 2006 yielded over 1800 applications, mostly from New Orleans and the surrounding parishes, of which about 1200 were eligible for the up to $45,000 in assistance. They were initially able to fund almost 300 and began distribution of the funds two months later (Fowlkes, 2008).

The application was fairly simple, requiring documentation of ownership, a scope of work and cost, and an assurance that work to be done with grant funds was directly related to damage caused by the storm. Houses that qualified needed to be part of a historic district, be over 50 years old and still maintain the architectural integrity that would contribute to the neighbourhood's historic designation. The grants were primarily to repair and replace those elements in homes that visibly designate it as a historic property (wood flooring, decorative windows, bracketed cornices, tile roofs, interior plaster, etc.).

A panel of five to seven people, including historians, preservationists and social workers ranked the applications, giving priority to owner-occupied properties and financial need as well as perceived viability of the proposed project. While there were a few plantation houses and notable structures given funding, most of the successful properties were typical small houses that contributed architecturally to the historic fabric of the neighbourhood. In some cases, however, properties that would have normally been given funding were so damaged that $45,000 would do little to bring the historic house to a stable condition.

The disbursement process was straightforward. The owner received 60 per cent of the grant at the beginning of the process and then was given two instalments of 20 per cent each at the middle and end of the scope of work promised. This grant was not deducted from a residents' RHP money nor was it deducted from insurance funds (most of the houses that received funding were un- or under-insured). SHPO did not wait to hear about a person's RHP funds before allocating grant funds and unlike those funds or money from FEMA, this money was not taxed.

By giving the citizens 60 per cent up front, SHPO demonstrated a level of trust in the citizens of New Orleans, as opposed to the attitude of the RHP. They also gave citizens a list of honest contractors that were capable of building to NPS standards, though owners were free to choose any licensed contractor. SHPO project officers visited each of their 600 properties regularly to answer questions and insure compliance of grantees and contractors.[12]

Part of the success of SHPO was its speed and they were given another 10 million dollars a year later, funding an additional 256 properties from the original grant pool. To date almost all the monies have been distributed and the houses funded have been properly renovated to historic rehabilitation guidelines established by the NPS. The applications were simple, the process transparent, and meaningful assistance and project management was provided along the way – a stark contrast to the RHP process administered by ICF under the auspices of the Louisiana Recovery Authority.

More Synergies 2: Foreign aid

Immediately after Katrina, offers of foreign aid came pouring in to Washington. From television images of people waving to be rescued from their roofs to the mass of humanity living in squalor in the New Orleans Convention Centre, America's poverty and racial inequity was broadcast around the globe. Other nations immediately came forward with offers of aid, both cash and in-kind. Of the $854 million offered by foreign governments to the US government, however, only $40 million has been claimed and used (Solomon and Hsu, 2007). While some aid was wasted due to the inability to coordinate and get the aid to the people in a timely manner, other funds were flat out refused. The Bush administration told other nations, in so many words: 'Thanks, but no thanks.'

These figures, however, do not tell the entire story. Some governments effectively gave their donations to outside groups, such as the Red Cross, instead. The government of Qatar made another decision: to give $100 million to help rebuild the Gulf Coast after Katrina via the establishment of the Qatar Katrina Relief Fund. The Fund provided direct cash assistance to groups and institutions to rebuild homes, to provide medical care and to pay for student scholarships to local colleges and universities.

Qatar's Ambassador to the US led the effort to identify local organizations that could efficiently and effectively distribute the funds. He solicited advice from governors, mayors and the congressional delegations of the affected areas as well as forming a distinguished advisory committee including community leaders. The success of this careful selection strategy is evidenced in the success of one of its grantees: The Qatar Tremé Renewal Project.[13]

The Qatar Tremé Renewal Project: A successful example of synergies and networks

For over 20 years the Covenant House New Orleans (CHNO), located in Tremé, has been providing comprehensive services to homeless and at-risk youth. Their secondary mission was the stability and preservation of the family, and they offered help to the community towards that goal. They were concerned that many families were not moving back or moving into unsafe conditions, such as improper plumbing or leaky roofs, or onto streets with many vacant houses. There was also a looming deadline set by the city council for owners to complete gutting and remediation lest their property be declared blighted and seized by the city and sold at auction. Given that Tremé is located on high ground adjacent to the French Quarter, the developers were circling.

The CHNO proposed a project to rehabilitate 100 homes of low- and middle-income families in Tremé. They envisioned helping long-time residents, key community leaders, first responders and teachers re-inhabit their homes towards stabilizing the community and encouraging others to return home. Their application was accepted by the Qatar Katrina Relief Fund and they were given $2.5 million to complete the project. At that point the CHNO became concerned that home building may be outside of its area of expertise and primary mission of helping the homeless and at-risk youth. There were also liability issues to consider. In early 2007 they formed the Qatar Tremé Renewal Project (QTRP) as a separate entity to carry out the grant.

In May 2007 they hired a project director, Cathy Puett, a licensed social worker with a background in non-profit organizations (Figure 7.9). While having no construction experience, Puett had the ability to build trust in the community and a practical notion of how to get things done. The QTRP developed a simple application for homeowners, primarily to determine if they: (1) had lived in the house prior to Katrina; (2) were within the project's income guidelines; and (3) were unable to re-inhabit their homes without outside

Source: Photo by I. Maret

Figure 7.9 *Puett at a progress meeting with a Tremé homeowner*

funds. Applications were vetted by a selection committee which was a heterogeneous group including a school principal, representatives of the CHNO board and other local NGOs, an attorney, an accountant and citizens from Tremé active in neighbourhood groups.

Construction began on the first homes two months later in July, lightening speed by post-Katrina New Orleans standards. The residents of Tremé welcomed Qatar's generosity and many felt that non-profit groups were actually driving the repatriation effort in comparison to the poor performance of federal agencies. They had on average about $25,000 per home to spend and the first few homes were a learning experience for Puett, allowing her to tighten the process along the way.

In the beginning it was difficult balancing all the variables of renovating storm damaged homes, particularly the coordination with other NGOs as well as with the homeowners themselves doing work on the house at the same time. Since QTRP had such limited funds they chose to fund a scope of work that included basic services and stabilization such as plumbing, electric and levelling the house. Other granting agencies such as the SHPO focused on the specific historic elements of the house, notably those that would keep the structure contributing to the Historic District status of the neighbourhood (Figure 7.10).

Source: Photo by I. Maret

Figure 7.10 *Dumee-Brasco House (c. 1845), now owned by a local schoolteacher and her family, is being repaired thanks to both the QTRP and the SHPO*

And yet other NGOs, such as Rebuilding Together, relied on large groups of volunteers from out-of-town and they typically did construction that lay people with some expert help could do: painting, laying flooring, some carpentry and dry walling. The combination of rebuilding funds and volunteers along with the owners themselves to bring a home to habitable status was a challenge for any project manager.

The process from application to completion was straightforward but offered some unexpected benefits to the homeowners and the community. First the homeowner completed a simple one-page application that was sent to the selection committee. After careful examination, a yes/no decision was made. Next, one of QTRP's two part-time construction managers wrote a scope document, specifying the work for bid. Puett then walked through the scope of work with the homeowner as some trade-offs and material choices may need to be made. Next the scope of work was submitted by QTRP to the Historic District Landmarks Commission (HDLC) for preliminary approval, as Tremé is a National Register Historic District. A licensed, insured contractor was selected among bids by the QTRP and they were responsible for obtaining building permits and HDLC compliance documents. The contractors are paid by QTRP at the end of every month for work completed.

The fact that the QTRP managed and paid the contractors is important. New Orleans had been fraught with contractor incompetence and fraud, particularly in the year following Katrina, and homeowners felt secure that the QTRP process had properly identified honest, qualified contractors. Many homeowners went on to negotiate for additional services with the same contractor and they helped their neighbours do the same.

As of 2010 most of the houses were completed or well into the process of completion, an amazing feat by a small non-profit. The speed and energy of the QTRP and their ability to garner community trust and support is commendable. They carefully chose the houses to rebuild in the hopes that they would bring with them many former long-term residents and, in turn, other residents will follow. The money spent by the QTRP was a 'soft second mortgage', that was forgiven after the owner lived in the house for three years. By making the grant a mortgage, the homeowner was not penalized by the federal and state government's official 'Road Home' programme when they finally got their money.

Conclusion: What STS has to offer disaster and repatriation research

In analysing the various successes and tensions in the rebuilding of Tremé several strategies stand out that are illuminated through theories from Science and Technology Studies (STS), namely Public Understanding of Science (PUS) and New Political Sociology of Science (NPSS). First is the issue of trust that the neighbourhood has in the specific organization, based on past interactions and reputation. Irwin et al (1996) has shown that trust is a primary indicator as to whether scientific and technical information disseminated by the organization is believable. Extrapolated to technologies of rebuilding, the issue is whether the NGO is believable in its intention to help the community rebuild, particularly given the pressure of outside gentrification and development in Tremé. For example, the issue of trust played out in a negative way in the city's attempt to force an outside charter school operator on the community, in part, by denying an open and participatory process in the decision-making process (Carr, 2010). Similarly, the closing of the Lafitte Housing Project was cloaked in secrecy, with opaque, top–down, expert-centred, processes driving decisions that adversely impacted the neighbourhoods ability to repatriate its former poorest residents. And looming large in a negative way, was the state's apparent lack of trust in its citizens in administering the Road Home Programme, dispersing funds in a paternalistic fashion, crippling the hope of many for a speedy return home. Trust is a reciprocal exchange and unfortunately in the case of many governmental agencies in post-Katrina New Orleans, was institutionalized as distrust in its citizens. Conversely, NGOs and SHPO engendered the trust of many citizens. Not only did the citizens trust the information and intention of these groups, they had demonstrated their trustworthiness even before the storm. Rebuilding Together, the preservationists, and Covenant House (to administer the Qatar project) all had worked *with* the community before Katrina and had track records of improving the housing and social conditions in Tremé.

Additionally, following PUS, the NGOs and SHPO demonstrated an openness to citizen participation in decisions about, and techniques of rebuilding. Rebuilding Together was founded on the idea that non-skilled, volunteer citizens and construction experts could effectively work side by side in renovating homes after the flood. Homeowners were also welcome, if available and able. In all cases homeowners were consulted on their ideas and wishes for their refurbished home. The QTRP and SHPO incorporated technical hybridity in a different manner. Both openly encouraged homeowners to continue their own renovation projects in tandem with professional contractors. Even though, at times, it was difficult to ascertain where the agency/NGO project stopped and the homeowners' project began, both groups felt that owner-driven rebuilding was key, even if a bit chaotic to keep track of. Inclusive processes were privileged over bureaucratic control.

All building help groups also adopted rules and regulations that enabled timely rebuilding without burdensome penalties for residents. Using the NPSS lens, these rules allowed for citizens to better access building supplies and contractors in contrast to the state Road Home Programme that coalesced power and access at a central point, frustratingly out of communication with most homeowners. These groups make apparent how institutional rules or technologies of operation impact community access to rebuilding and significantly condition the distributive aspects of community reconstruction. Democratic-oriented policies and their inclusive synergistic institutional arrangements greatly enhanced the timeliness and effectiveness of repatriation help organizations.

As of July 2009 Orleans parish has a population of about 350,000 that is about 100,000 less than the pre-Katrina number (Greater New Orleans

Community Data Centre, 2010b). This number includes not only former residents but also those who have moved to the city for construction jobs or those who came to clean up after the storm and decided to stay. The demographics are also changing, notably the parish is less African-American (60 per cent compared to 67 per cent before the storm) and more Latino (4.5 per cent compared to 3 per cent before) and more white (31 per cent compared to 27 per cent) in the post-Katrina era (Greater New Orleans Community Data Centre, 2010a).The fate of many other former residents from more damaged neighbourhoods remains uncertain. As the final Road Home Programme money is distributed and NGOs gain momentum from hindsight, we are hopeful that repatriation will continue. From this research it is clear that more than money is needed for rebuilding neighbourhoods; trustworthy organizations with democratic policies that are willing to work together and in concert with community members are necessary for expeditious rebuilding and citizen return.

Acknowledgements

This research was assisted by grants from: the Katrina Program of the Social Science Research Council with funds provided by the Russell Sage Foundation, Virginia Tech's Institute for Society, Culture and Environment, the Bogliasco Foundation, and the NSF grant # SES-0821353. The views represented in the chapter are our own.

Notes

1 In the 1960s, the decision to route the new highway through Tremé was considered to be a victory for many historic preservationists. The original plan, developed by the infamous Robert Moses, was to have an elevated expressway along the river, dividing the French Quarter from the levee.
2 There is also evidence that at least one tornado touched down after the storm, wiping out entire buildings along a linear path through the community.
3 Interview with Jerome Smith, 18 October 2007, in New Orleans at the Tremé Community Centre, by authors.
4 Interview with Jerome Smith, 18 October 2007, in New Orleans at the Tremé Community Centre, by authors.
5 While gentrification was happening slowly before Katrina, it was hastened after the flood due to the fact that most of lower Tremé did not have water as it is on high ground. Additionally, it is located adjacent to the French Quarter, a highly desirable locale in the city.
6 Interview with Louis Charbonnet, 16 October 2007, at his funeral home in New Orleans, by authors.
7 Smith, interview by authors.
8 According to internal emails and memos, the cost of renovation was approximately \$85,000,000 whereas rebuilding was estimated at \$100,000,000. See Reckdahl (2006a).
9 See Hammer (2008). The article also notes that citizens with high value properties lost more actual dollars in the RHP scheme as their homes were valued well above the programme's limit.

10 They are allied with the national group Rebuilding Together, Inc., headquartered in Washington DC.
11 From an interview with Kristin Gisleson Palmer, Director, Rebuilding Together New Orleans, 7 December 2007, by Barbara Allen.
12 Interview with Dan DiGiuseppe, Sr Project Officer and Tracy Nelson, Program Supervisor, Historic Building Recovery Grant Program, 10 December 2007, by authors.
13 The original name of the group was The Qatar Tremé/Lafitte Renewal Project, but the Lafitte Public Housing Project became a very contentious political issue and the group felt its work could best be accomplished by concentrating on the houses in the Tremé neighbourhood.

References

Allen, B. (2007a) 'On performative regionalism', in V. Canizaro (ed) *Architectural Regionalism*, Princeton Architectural Press, New York, pp417–426

Allen, B. (2007b) 'Environmental justice, local knowledge, and after disaster planning in New Orleans', *Technology in Society*, vol 29, pp153–159

Allen, B. (2007c) 'Environmental justice and expert knowledge in the wake of a disaster', *Social Studies of Science*, vol 37, no 1, pp103–110

American Medical Association, 'Post-Katrina death rate in New Orleans shows significant increase', www.ama-assn.org/ama/pub/category/print/17710.html, accessed 17 June 2008

Carr, S. (2010) 'Debate over charter operator for Craig Elementary raises questions about neighbourhood's role in remaking local schools', *The Times-Picayune*, 7 March

Crutcher, M. E. Jr (2001) *Protecting Place in African American Neighbourhoods: Urban Public Space, Privatization, and Protest in Louis Armstrong Park and the Tremé, New Orleans*, PhD dissertation, Louisiana State University, UMI Microform 3016535

Eden, R. and Boren, P. (2008) 'Timely assistance: Evaluating the speed of Road Home grantmaking', Gulf States Policy Institute, RAND Corporation, www.rand.org/pubs/documented_briefings/2008/RAND_DB557.pdf, accessed 10 December 2008

Fischer, F. (2000) *Citizens, Experts, and the Environment*, Duke University Press, Durham, NC

Foster, M. (2008) 'In New Orleans, demolition begins on public housing', www.preservationnation.org/magazine, accessed 9 April 2008

Fowlkes, S. (2008) 'Lieutenant Governor's Historic Building Recovery Grant Program achieves goals', *Preservation in Print*, vol 35, no 7, p7

Frickel, S. and Moore, K. (eds) (2006) *The New Political Sociology of Science: Institutions, Networks, and Power*, University of Wisconsin Press, Madison, WI

Greater New Orleans Community Data Centre (2010a) www.gnocdc.org, accessed 10 June 2010

Greater New Orleans Community Data Centre (2010b) *Census Population Estimates 2000–2009 for New Orleans MSA*, www.gnocdc.org/census_pop_estimates.html, accessed 10 May 2010

Hammer, D. (2007) 'Road Home appeals scarce; critics say process shrouded in secrecy', *New Orleans Times Picayune*, 2 March 2008

Hammer, D. (2008) 'Did Road Home treat all neighbourhoods fairly?', *New Orleans Time Picayune*, 27 May

Irwin, A. (1995) *Citizen Science*, Routledge, New York

Irwin, A. and Michael, M. (2003) *Science, Social Theory and Public Knowledge*, Open University Press, Philadelphia, PA

Irwin, A. and Wynne, B. (eds) (1996) *Misunderstanding Science?*, Cambridge University Press, New York

Irwin, A., Dale, A. and Smith, D. (1996) 'Science and hell's kitchen: The local understanding of hazard issues', in A. Irwin and B. Wynne (eds) *Misunderstanding Science?*, Cambridge University Press, New York, pp47–64

Johnson, J. (1991) 'New Orleans' Congo Square: An urban setting for early Afro-American culture formation', *Louisiana History*, vol 2, no 2, pp117–157

Lovett, J. (2006) 'Why can't our Road Home be as simple as Mississippi's?', *New Orleans Times Picayune*, 17 August

Providence Community Housing (undated) 'Providence Community Housing', www.providencecommunityhousing.org/2010-06%20Plans_and_Progress.pdf, accessed 9 April 2008

Rebuilding Together New Orleans (2010) *About Us*, www.rtno.org/about-us, accessed 10 May 2010

Reckdahl, K. (2006a) 'Like a ton of bricks', *Gambit Weekly*, 24 October, www.bestofneworleans.com/dispatch/2006-10-24/news_feat.php, accessed 9 April 2008

Reckdahl, K. (2006b) 'Razing a community', *Gambit Weekly*, 31 October

Reckdahl, K. (2007) 'Culture, change collide in Tremé', *New Orleans Times Picayune*, 2 October

Road Home Program (2008) *Situation and Pipeline Reports*, www.road2la.org/newsroom/pipeline.htm, accessed 10 September 2008

Simon, D. (2007) 'Damage closing Tremé school', *New Orleans Times Picayune*, 15 November

Solomon, J. and Hsu, S. (2007) 'Most Katrina aid from overseas went unclaimed', *The Washington Post*, 29 April, ppA01

Spitzer, N. (2006) 'Rebuilding the "Land of Dreams" expressive culture and New Orleans' authentic future', *Southern Spaces*, 29 August, www.southernspaces.org/contents/2006/spitzer/1a.htm, accessed 10 September 2006

Sturgis, S. (2006) 'Under growing fire, Louisiana's Road Home Program makes changes', http://southernstudies.org/facingsouth/2006/12/under-growing-fire-louisianas-road.asp, accessed 5 June 2008

Warner, C. (2007) 'Road Home eyeing Mississippi program', *New Orleans Times Picayune*, 26 March

Part III
International Disasters and Katrina Comparisons

8
The 2002 Flood Disaster in the Elbe Region, Germany: A Lack of Context-sensitive Knowledge

Juergen Weichselgartner and Emilie Brévière

Introduction

The Elbe River basin covers different geographical regions from middle mountain ranges in the west and south to large flatlands and lowlands in the central, north and eastern part of the basin. From its 1394m high origin in the Czech Bohemian mountains to its mouth in the German City of Cuxhaven at the North Sea, the Elbe River has a length of 1165km (700km in Germany) and a drainage basin of 148,268 sq km (two-thirds in Germany) (BfG, 2002, p1). The Elbe region is the driest of the five largest German river basins, resulting in frequent water stress and water deficiencies. The summer of 2003 brought a severe drought and water deficit only one year after the great floods.

Starting on 6 August 2002, a complex weather situation over central Europe led to heavy and widespread precipitation in Hungary, Austria and south-western parts of the Czech Republic, as well as eastern and southern Germany. Coming from the British Isles, the low pressure system *Ilse* passed along the southern part of the Alps and then headed for Saxony in a so-called Vb track. In a ten-day period from 1 to 10 August, precipitation of about 60mm fell widely in the Elbe River drainage basin area, followed by strong rainfall from 11 to 13 August (BfG, 2002, p9). During this period, many areas in eastern Germany had rainfalls of 100–200mm/day, some more than 300mm/day: a precipitation in three days that was two to four times as much as the usual monthly average (LfUG, 2004).

The unusual intensity of rain exceeded the soil's capacity to retain water and generated two types of floods in the Elbe river basin: flash floods at the tributaries of mountainous catchments (e.g. Weißeritz River, Upper Mulde

Table 8.1 *Historical water levels compared to 2002 at different gauge stations of the Elbe River*

Gauge station	*River km*	*Highest water level*		*Water level 2002*
		Date	*cm*	*cm*
Usti n. L.	–38.70	not available	1119	**1185**
Schöna	2.10	April 1941	868	**1202**
Dresden	55.60	March 1845	877	**940**
Riesa	108.40	March 1988	755	**945**
Torgau	154.60	March 1940	863	**945**
Wittenberg	214.10	November 1980	624	**707**
Aken	274.80	March 1845	740	**765**
Barby	295.50	March 1845	733	700
Magdeburg	326.60	February 1941	701	670
Tangermünde	388.20	March 1981	670	**767**
Wittenberge	454.80	January 1920	715	**734**
Schnackenburg	474.60	April 1988	692	**751**
Dömitz	504.70	March 1888	744	657
Neu Darchau	536.40	April 1895	724	**732**

Notes: Point Zero is the Czech-German border. Figures in **bold**: new records in cm
Source: Modified from BfG (2002, p32)

River) with high discharge dynamic and flow velocity, and slow swell floods of the Elbe River and the lower reach of its main tributaries (e.g. Vltava River, Joint Mulde River). In the Saxon capital of Dresden, the Elbe River rose from a normal summer level of about 2m, surpassing the historical flood mark of 8.77m seen in March 1845, to reach a water level of 9.40m – the highest level ever recorded (Table 8.1).

Flood losses

Both the Elbe River floods and Hurricane Katrina exposed the vulnerability of highly developed countries to external shocks and brought national awareness of the deficiencies in preparedness for and response to disasters. Their impacts on urban areas turned both events into one of the costliest disasters and disaster relief in the history of the respective countries (Mechler and Weichselgartner, 2003; Cutter et al, 2006). In Germany, the floods caused 20 fatalities and directly affected 337,000 people in eight federal states (*Länder*) (DKKV, 2004, p8). Shortly after the floods, official sources issued damage estimates of €25–30 billion, which, in a press release from the Federal Ministry of Justice in December 2002, was revised to around €9.1 billion (Bundesministerium der Justiz, 2002). For comparison, Kates et al (2006) estimated that Katrina accounted for 1570 deaths of Louisiana residents and caused an aggregate monetary loss of around $40–50 billion in Orleans Parish including direct property losses ($20–22 billion), still ongoing economic losses ($4–8 billion), and emergency assistance ($16–20 billion).

With two-thirds of the total losses, the federal state of Saxony was hit hardest. The losses of approximately €6 billion made up around 40 per cent of the annual Saxon government budget. Considered spatially, losses were most severe in the region of Dresden, where more than half of the losses occurred. In the Saxon capital, the floods caused €400 million to public assets, €395 million to private households and €67 million to businesses (according to government officials, roughly 25 per cent of the companies had flood insurance). Many cultural landmarks were damaged, including the Semper Opera, the Zwinger Palace museum and the train station. Viewed by sector, the major share of more than one-third accrued to federal state-owned and municipally owned infrastructure, followed by private households and private companies. Facing losses of almost €800 million, Allianz Insurance Group was hit particularly hard because the home insurance policies that it had taken over in eastern Germany during the reunification process provided blanket coverage for flood damage. Much of the damage was caused in small areas by tributaries of the Elbe River.

Loss compensation

The way the losses were compensated illustrates how different responsibility can be perceived. More than four years after Hurricane Katrina, federal judge Stanwood Duval accused the Army Corps of Engineers of 'insouciance ... and short-sightedness' (Pilkington, 2009). The judge said it had 'myriad' ways in which it could have acted to prevent disaster but failed to do so and found that the Corps, which was responsible for maintaining the waterways and levees that protected New Orleans from flooding, had known of the threat to areas of the city and yet had failed to act in time: 'The Corps' lassitude and failure to fulfil its duties resulted in a catastrophic loss of human life and property in unprecedented proportions.' The ruling opens the door to further claims from up to 100,000 victims with possible settlements running into billions of US dollars in compensation claims.

In contrast, the Elbe floods were considered a case for national solidarity rather than of individual responsibility and, therefore, the Federal Government, the federal states and the European Union were instantly willing to fund an unprecedented portion of the damages. The declared will of the national government was that 'nobody should be worse off than before the floods'. National solidarity was invoked and in view of the extreme flood and the severe damages, the Federal Government and the federal states immediately set up a special disaster relief and reconstruction fund to finance the flood losses – the so-called *Sonderfonds Aufbauhilfe* amounting to €7.1 billion (Mechler and Weichselgartner, 2003).

The financing measures fell into two phases. In the emergency relief phase, unbureaucratic, immediate assistance was provided and private households were compensated with €500/person (max. €2000/household), residential property with €5000/building, business with €15,000 and €500/employee respectively, and agriculture and forestry with €50,000. In the reconstruction assistance phase, more rules were employed and a more careful evaluation was undertaken in order to allocate money to those in greatest need. The infrastructure sector was allotted the highest compensation ratio with funding for 90 per cent of the reconstruction costs. Maximum reconstruction costs covered for private households were set at 80 per cent, while for businesses they were set at 75 per cent. The compensation limit for agriculture and forestry amounted to €1 million or 30 per cent of crop losses. Insurance disbursements were regulated in order to prevent overcompensation.

The provision of government funds to the affected private households, companies and municipalities was governed by a set of principles to guarantee the efficient and fair allocation of the funds, allow quick reconstruction and provide incentives for *ex ante* measures. These principles included *subsidiarity* (the delegation of responsibilities to the lowest administrative level feasible), *parallelity* (reconstruction in the affected East German region was to run parallel and independent of the reconstruction in East Germany after reunification), *provision of incentives* (inclusion of deductibles in order to maintain incentives for mitigation and insurance), *efficiency* (financing of direct losses only to primarily compensate those worst affected), and the focus on the *ability* to rebuild (loss financing was provided in terms of reconstruction costs rather than current values).

In addition to governmental funding, insurance coverage (estimated to amount to €1.8 billion), funds from the European Union (€444 million) and voluntary public donations played an important role in loss compensation. For instance, the German Red Cross Society (DRK) initiated a 'Neighbours in Need: Flood 2002' project and the national soccer team donated around €500,000 to the flood victims. On 16 August, a fundraising gala collected €16 million – the largest donation ever collected in German television history. The media activity 'Hamburg helps floods victims' of former chancellor Helmut Schmidt involved nine newspapers and TV channels and collected almost €13 million. In total, the general public donated more than €242 million.

Historical-cultural context

A look at the past shows that there had been comparable extreme flooding in the basin of the Elbe River on previous occasions (Figure 8.1). More than 200 years ago, the authorities in Saxony had already successfully stemmed major flood risks, after extensive damage resulting from an extreme flood event in February 1784 (Weichselgartner and Sendzimir, 2004, p6). In the winter of 1798/99, realizing that the weather pattern might create dangerous flood conditions around the Elbe River, the authorities designed an early warning system based on collective close observation of the environment and the

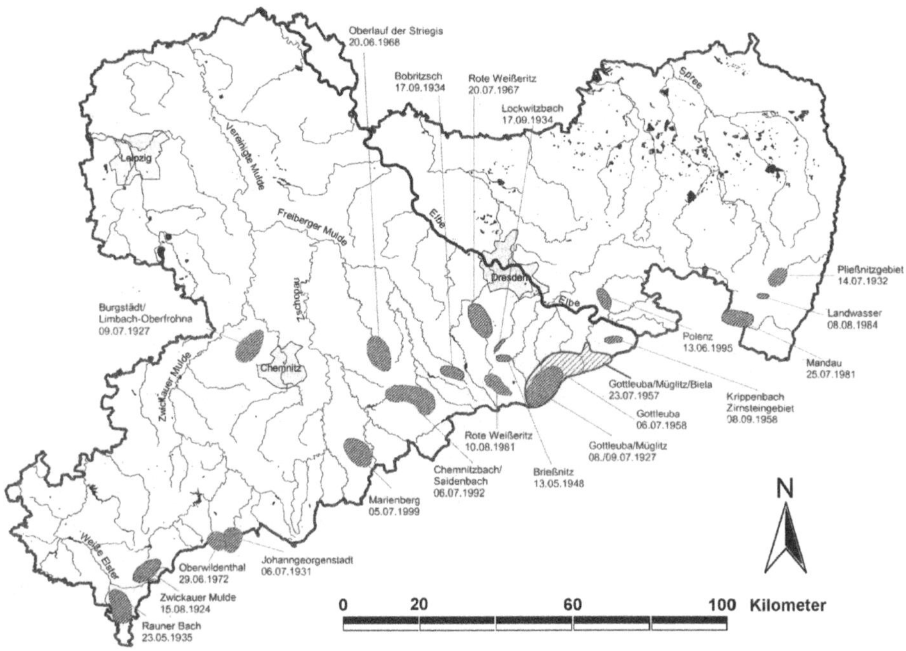

Source: Modified from SMUL (2002)

Figure 8.1 *20th century local heavy summer rainstorms with disastrous impacts in Saxony*

involvement of the population and the military. Sixteen signal cannons were implemented along the Elbe River and a legal act was passed to regulate flood prevention measures.

While the damaging force of rivers has always been a constant in nature, the societal perception and the interpretation of the natural forces has changed over time, shifting from supernatural or mystical forces, to nature and, with some reluctance, to humans, or to society as a whole, who have made 'incorrect' decisions (Weichselgartner and Bertens, 2004). Until the 18th century, floods were interpreted as 'acts of God', which paralysed scientific arguments and hindered technical prevention measures. Later on, they became 'technical problems' posed by nature, for which engineers have to find solutions. Likewise, disaster response developed from rather unplanned efforts to aid and care for disaster victims, for example, the Catholic Church in Germany, to a network of laws and regulations. Over the course of the late 19th and 20th centuries, these laws and regulations supplemented and increasingly replaced private charitable efforts with state-administered programmes. Viewed temporally, it becomes evident that the way people think about natural disasters influences not only where they look for solutions, but also the shape and character of the means they used to attain these solutions.

In addition to the changes over time, a society's responses depend on economic and social institutions and on its cultural values (for an illustrative description see Diamond, 2005) and thus geographical differences in viewing and managing natural disasters exist. Despite comparable economic, social and cultural history and structures, the countries in Western Europe operate their flood response and crisis management in different ways (Rosenthal and 't Hart, 1998). Today, spatial differences can be observed with regard to climate change adaptation processes as well (Roggema, 2009). Cultural diversities also exist between Germany and the United States. They might be related to the fact that environmental issues are understood quite differently: 'A comparison of developments and the various arguments deployed in the debate is of interest because America's founding myth is based upon the glorification of the frontiersman who endures nature's trials, survives without outside help, and derives his identity and strength largely from these experiences' (Lübken and Mauch, 2004, p104). The organizational culture lens applied by Dowty et al (this volume) to examine disaster response shows that there are social-cultural differences regarding the search for responsibilities, blame and future planning strategies. Such variations are not solid constants but variable. For instance, the 9/11 terrorist attacks have reintroduced a focus on emergency management (away from mitigation) and a framework in which people are mostly perceived as passive victims rather than being part of sets of unequal access to opportunities and unequal exposures to risks which are a consequence of societal processes.

Both Katrina and the Elbe River floods have revealed that people approach their risk-conditions beyond the operational disaster paradigm in that they do not perceive disaster issues as separate phases of a disaster management cycle, but as integral parts of their coping strategy and development process. Although it is beyond the scope of this chapter to offer a detailed social-cultural comparison (see Mauch and Pfister, 2009), it is crucial that disaster research enhances its temporal depth with studies addressing long-term structural questions and cultural issues. Positive examples have shown that natural disasters can result in fundamental learning processes and in improvements in public and private institutions (Poliwoda, 2007).

Flood risk management

With regards to flood risk management, two main mitigation instruments are commonly applied in Germany and in the United States: (1) structural measures, for example, reinforcing buildings, creating flood levees, retention areas and dikes, and (2) non-structural measures, for example, restrictions, hazard zoning, monitoring, forecasting, early warning systems, incentives and information management. In both 'soft' and 'hard' measures, different levels of spatial scales are involved, ranging from allotments (building laws) to communities (retention areas), states (disaster protection) and national levels (navigation).

In the Elbe River basin and its tributaries, there are several monitoring activities coordinated by the International Commission for the Protection of

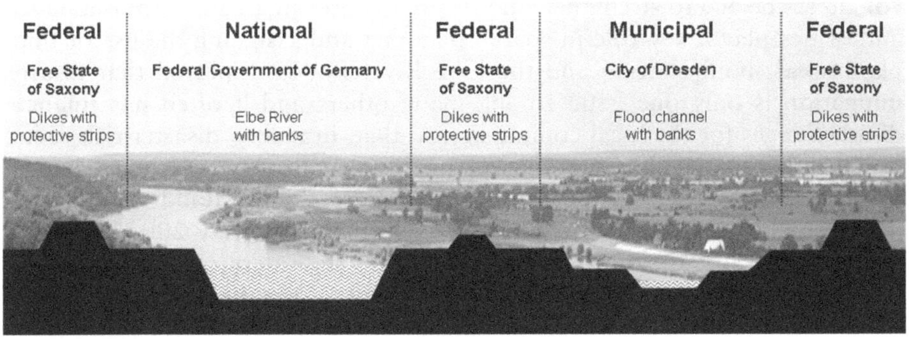

Source: Modified from DKKV (2004, p14)

Figure 8.2 *Responsibilities in Dresden for the federal waterway Elbe River*

the Elbe (the contracting parties are Germany, the Czech Republic and the European Union). In Germany, the 'ARGE Elbe' (*Arbeitsgemeinschaft für die Reinhaltung der Elbe*) – a working group comprised of the federal states of Hamburg, Lower Saxony, Schleswig-Holstein, Saxony, Saxony-Anhalt, Mecklenburg-West Pomerania and Brandenburg – is in charge of monitoring the Elbe River. Through the cooperative efforts of river basin commissions at different governmental levels and formulated programmes of action, a wide range of flood protection activities exist. However, the 'unbalanced coexistence' of mitigation measures, often not developed across sectors, was a major drawback in 2002. Figure 8.2 illustrates the division of responsibilities for the Elbe River in Dresden.

Bowden (this volume) highlighted the unexpected connections between natural, technological and social systems. In Saxony in 2002, the Elbe River floods similarly disclosed the variety and complexity of such connections governing the management of water bodies. A lack of understanding with regard to these connections resulted in inadequate cooperation, inadequate communication and inadequate management across administrative and spatial boundaries. Ultimately, diverse and non-coordinated spatial planning and flood protection activities along river catchments and river reaches condition heterogeneity within the technological system. For instance, along the Elbe River, around 127km of dykes provided protection for floods with a 100-year recurrence interval. However, 93km of these dykes were built before 1900 and 32km were constructed between 1900 and 1945, resulting in inhomogeneous materials and structures with a decrease in reliability. Construction practices also played a role in New Orleans. Instead of overtopping, water flowed through the collapsed earthen levees – ones that had been upgraded with a thick concrete wall.

But unlike Katrina and the atypical disaster responses in New Orleans (see Bowden this volume), the Elbe River floods showed socially unifying effects. Beside the already mentioned high solidarity after the floods, thousands of volunteers helped to strengthen and secure the breaking dams. Although local authorities play a key role in spatial planning and assigning the use of flood plain areas, both Katrina and the Elbe River floods displayed that disaster mitigation is only one issue among many others and it often has financial disadvantages for the local community, so that, in reality, disaster-risk reduction through spatial planning still remains a 'strong instrument in weak hands'.

Last but not least, due to the Elbe River floods, flood and disaster management have changed over recent years. Since 1 May 2004, Germany has a central organization working to ensure the safety of the population, combining and providing all relevant tasks and information in a single place: the Federal Office of Civil Protection and Disaster Assistance (*Bundesamt für Bevölkerungsschutz und Katastrophenhilfe*, BBK). In cases of emergency, the federal government controls the disaster relief and civil protection programmes. The Federal Agency for Technical Relief (*Technisches Hilfswerk*, THW) and the local fire departments are part of these programmes. If necessary, the German Armed Forces (*Bundeswehr*), the German Federal Police (*Bundesgrenzschutz*) and the 16 state police forces will also be deployed for disaster relief operations. The Federal Government adopted a five-point programme to improve preventive flood protection and a new protective law for floods, requiring the federal states to define flood plain areas for all high flood risk riverine areas by 2012 (see Box 8.1).

Flooding scenarios and early warning

In Germany, flooding scenarios are determined for most rivers and coastal areas, using design and other frequent discharges. Usually, the first prognosis can be made a few days before the peak levels will be reached. For the last decade, it has also been common to account for higher discharges than the design discharge, so-called extreme flood events. For the simulation of flood scenarios, observed, hypothetical and, increasingly, the forecast data of the Intergovernmental Panel on Climate Change (IPCC) are used. The German Armed Forces regularly use such scenarios in their exercises.

The federal states are responsible for flood warning. The warning systems are used only 6–72 hours in advance to avoid false alarms. At the federal level, state data are brought together and used for information and warnings. Represented on all significant federal navigable waters including rivers, canals, locks and seaward approaches within the sovereign territory of Germany, the Federal Waterways and Shipping Administration (*Wasser- und Schifffahrtsverwaltung des Bundes*, WSV) is responsible for flood forecasting in shipping routes, such as the Elbe River. The WSV is organized on three administrative levels: (1) supreme authority: the Federal Ministry of Transport, Building and Urban Development (*Bundesministerium für Verkehr, Bau und Stadtentwicklung*, BMVBS); (2) intermediate authorities: seven Waterways and Shipping Directorates; and (3) local authorities: 36 Waterways and Shipping Boards. Furthermore, various authorities are responsible for special tasks and duties, such as naval construction, hydrology, marine casualties investigation, combat

Box 8.1 Act to Improve Preventive Flood Control

On 10 May 2005, the new Flood Control Act (*Gesetz zur Verbesserung des vorbeugenden Hochwasserschutzes*) entered into force. For the first time, the Act to Improve Preventive Flood Control lays down uniform and stringent legal provisions for the prevention of flood damage on a nationwide level. Among others provisions, the federal states are obliged to designate more areas as flood plains than before. Waters and water segments have to be identified, along with flood plains, and designated according to the risk of flood damage. The federal states are obliged to inform the general public about their decisions. For areas with a high potential for damage, flood plains have to be designated within five years; the flood plains in flood-prone areas must be designated within seven years. The public has to be included in this process. The basis for designating flood plains is the 100-year flood level.

In future, planning new housing areas in flood plains is prohibited by federal law. No new buildings may be planned in these areas. Exceptions are possible if nine closely defined requirements are met, all of which have to be fulfilled completely in every individual case. They include that the municipality concerned has no alternative for human settlement development, that no lives are at risk, no significant property damage is to be expected and that the structure of new buildings is adapted to floods. Moreover, oil heating systems in flood plains have to be constructed in a flood-proof way or they have to be retrofitted. The installation of new oil heating systems may be prohibited if possible damages are too severe. In the case of agricultural land use, the federal states must take care to ensure that soil erosion and the pollution of water bodies are prevented or reduced.

If no flood protection plans exist, the federal states have to draw up plans coordinating flood protection along the rivers within four years. These plans have to be based on a 100-year flood. In addition to flood plains, the federal states have to designate flood-prone zones. Flood plains and flood-prone zones have to be marked in spatial plans, land use plans and development plans.

of oil and harmful substances pollution. The authorities of the federal states operate the flood forecasting for all other rivers within their state. The flood forecasting systems for the German coastal areas is operated by the Federal Maritime and Hydrographic Agency (*Bundesamt für Seeschifffahrt und Hydrographie*, BSH).

In general, the early warning procedure follows predefined steps and alert levels. In the case of a severe threat due to flooding, a crisis management group (*Krisenstab*) will be assembled. This group consists of experts from different fields, such as disaster management authorities, fire departments, water boards and relief organizations. In 2002, ten centres were responsible for flood forecasting on the Elbe River: nine *Länder* institutions and one federal office (BfG, 2002, p45). Flood forecasts were made on a daily basis at the Flood Forecasting Centre Elbe in Magdeburg using the forecasting system ELBA and data from 11 gauge stations at the Elbe, 3 at the Saale River and

1 at the Havel River. Despite the fact that numerous gauge stations (namely: Seidewitz, Müglitz, Rote Weißeritz, Wilde Weißeritz, Vereinigte Weißeritz, Vereinigte Mulde, Freiberger Mulde, Bobritzsch, Striegis) at tributaries were destroyed, flood forecasts were still possible in 2002. This was important since the forecasts served as a planning basis for the emergency operation centres. In contrast, no reliable, multimedia technical communication network existed, and radio and mobile telephones were mostly used.

Emergency planning and evacuation

Emergency planning is done on the federal state level and municipal level. Flood protection exercises for the proofing of emergency plans and for training purposes are common. In such exercises, not only crisis management software (the system itself and handling by the user) and technical flood protection devices are tested, but also communication channels and decision-making processes in uncertain situations. Nowadays, emergency plans are used in combination with web-based flood information and warning systems, containing various components such as current flood depth data, automated alarm and operation plans, integrated geographic information systems (GIS), two-dimensional flood models, the internal mailing system and resource administration. Furthermore, emergency management plans often include evacuation plans. In the past, evacuation has been carried out locally before and during flood events, but large-scale evacuations are not rehearsed physically in Germany.

In accordance with Germany's constitution, the Federal Government only has framework responsibility for flood defence policy and thus, the federal states are responsible for much of the implementation and for transposing the provisions into their own laws. In recent years, awareness has increased that past planning practices have resulted in the disconnection of rivers from floodplains with a resultant loss in biodiversity and a requirement for expensive structural flood defence measures. It was not until the Elbe River floods in 2002 that legislation was drafted, based on a preventative approach in which floodplains would be conserved by restricting activities such as construction and agriculture.

Likewise, Katrina may reignite interest in bolstering the wetlands south of New Orleans to provide more of a hurricane barrier. In 1998, a collection of state and federal agencies proposed 'Coast 2050', a $14 billion strategy to restore Louisiana's wetlands. But the project never won federal funding and hasn't moved beyond the planning stage. Thus, up to 100 square kilometres of the wetland buffer disappear each year, largely because the Mississippi River has been leveed and dammed so much that it deposits much less sediment onto the delta (Travis, 2005, p1657f).

Lessons learned

The extreme flood levels and high public solidarity led to an exceptional modus operandi of governmental response measures – the reconstruction fund as an 'externally driven measure', and the setting up of a special, independent commission as an 'internally driven measure'. This Kirchbach Commission (named after retired General Hans-Peter von Kirchbach who is well-known for his successful role in commanding the German Armed Forces during the Oder River flood in 1997) evaluated the response measures taken during the Elbe River floods and elaborated on proposals for modern disaster protection structures. The 252-page assessment mainly criticized the public administration and its communication structures by demonstrating shortcomings of the warning systems, stating that relevant information was available, but not being transmitted to the right places, or not in a useful fashion (Kirchbach et al, 2002).

In addition, the German Committee for Disaster Reduction (DKKV) – the national platform and coordinating point concerning disaster mitigation – initiated a process for evaluating the events surrounding the Elbe flood. The committee set up an interdisciplinary team of scientists to analyse organizational deficits and identify deficiencies in coordinating disaster prevention and response. After a survey on perception and adaptation, analysing the operational shortcomings, evaluating the numerous reports available on the Elbe floods and through supplementing these with their own surveys and analyses, the experts documented four structural deficits (DKKV, 2004):

1 Lack of connectivity between actors integrated in disaster defence: including an absence in relations in day-to-day activities, deficits in communication and a lacking of willingness to cooperate. This is mainly not a question of technical equipment but of soft skills.
2 Lack of common orientation: including organization-related thinking, disconnected developments and missing overview concerning the qualification and equipment of relief units. Everybody wants to have and provide everything, complementary equipment is not considered.
3 Weakness of disaster defence instances: including lack of knowledge, self-discipline, motivation, ability as well as over- and underestimation of own and other sources. Legitimacy of rules and regulations is not always voluntarily accepted and focus on the common good becomes lost.
4 Structural nucleus of the operative-tactical subsystems: including exclusion of innovations and controversies over power and, thus, obstructing the work flow. In addition to the structural deficits, consolidated findings were obtained concerning the reduction of damage potential and run-off, flood warning and disaster prevention.

Moreover, the lessons-learned study drew up a comprehensive picture of what should be taken into consideration for the future of flood risk reduction in Germany and presented strategies, measures and recommendations to reduce future flood losses (Box 8.2). Likewise, a new inter-agency performance evaluation task force provided answers to fundamental questions about the performance of the hurricane protection and flood damage reduction system in the New Orleans metropolitan area (Demirbilek, 2010).

Box 8.2 Selected recommendations for improved flood risk reduction

- The reduction of potential damage has top priority. Risk reduction through spatial planning has to be strengthened. In future, statutory regulations on keeping flood-plains undeveloped must be clearly defined. For better implementation of risk reduction through spatial planning, synergy effects between flood risk reduction and other interests, above all nature protection, landscape conservation or drinking water protection should be sought and used. Financial and insurance incentives are required for rebuilding in other places or for resettlements, in order to achieve a reduction of damage potential in the flood plains rather than just stagnation.
- Measures and recommendations must be comprehensively weighted in accordance with their importance for flood risk management, in particular, in accordance with the effectiveness of the measures vis-à-vis space, time and process intensity. Measures for evaluating effectiveness must be developed to be able to assess their importance. Measures for flood risk reduction and response must be assessed in accordance with specific catchment areas and with regard to their interaction. 'Flood protection concepts' must take into consideration the broad range of flood risk reduction measures.
- Limits to natural retention, in particular its limited effectiveness in the case of extreme discharges, must be recognized and accepted.
- Technical flood protection equipment is essential for reducing extreme flooding. However, their limitations and risks must be made transparent. Clear responsibilities must be created and the dimensioning of the equipment must be more closely related to the protection objectives. The equipment must be maintained: here the dikes, in particular, are not up to standard. There must be frank information about existing weak points.
- Possibilities and limits for taking action must be emphasized. The usual HQ-100 observations must also be supplemented by further scenarios as well as by cases of failure. Presentations of danger situations should be intensified and should take place permanently and be linked to precautionary strategies and response possibilities. Warning systems for specific dangers and regions, ranging from gathering data and forecasts right through to the reaction of affected persons should be expanded. Here, specific action must be incorporated in integrated concepts. Technical security and material and action-orientation of communication must be improved.
- To successfully implement protection concepts, a discussion process must be introduced that involves the whole of society and involves the whole population. Quick suppression of and forgetting about negative situations must be countered.
- Flood risk reduction and flood response are cross-sectional tasks and require a great deal of communication, cooperation and management. All participants from different specialist and spatial areas must be better integrated with each other. Interfaces are weak points, and must be regularly tested for functionality and then updated.
- Own private precautions, constructional, behavioural and (insurance-aided) risk reduction should be systematically developed and stimulated as a component of flood risk reduction. Own precautions are neither encouraged nor rewarded sufficiently. This depends a great deal on whether flood dangers and limitations are

recognized and whether possibilities for taking action are known. In the elementary damage insurance, insurance premiums and deductibles should be oriented not only to the risk, but also the extent of own precautions, for instance, to more intensely promote preventative building measures. To make sure that solidarity becomes more effective in individual cases, it must be ensured that private and commercial precautions can be integrated in (state) disaster response well before the disaster.

- Under the primate of flood protection, the interests of a broad range of political areas must be integrated in the drawing up of flood risk reduction concepts at an early stage. An interdisciplinary, space-oriented risk management is required, which demands a stronger integration of water (resources) management, land (resources) management and spatial planning.
- Action covering whole river catchment areas and extending across borders is essential for preventative flood risk reduction. Implementation of transnational programmes and network forming activities is suggested.
- Solidarity with subsequent generations requires decisions on flood risk reduction concepts despite great uncertainties. First, lessons must be learned from disasters and above all design reconstruction, so that a higher level of disaster reduction is achieved. Today's flood risk reduction should not be allowed to restrict future generations' possibilities for taking action. Flexible systems and expandable concepts are therefore required.

Source: Modified from DKKV (2004, p5f)

Barriers in the science–practice–policy interface

In both the US and Germany, an investigation of prevention and response processes was conducted, characterizing the event as 'a failure of initiative' rather than 'a failure of knowledge' (DKKV, 2004; Select Bipartisan Committee, 2006). But there is, of course, a nexus between the two: knowledge and initiative require information and a coordinated process for sharing it. Frickel and Vincent (this volume) have shown how the absence of knowledge shapes contentious politics and how contentious politics in turn shapes the non-production of knowledge. In addition – and this is the authors' main concern – both disasters revealed shortcomings in effectively transferring available theoretical knowledge into practical application. Katrina was known to be heading towards New Orleans a few days in advance, and the potential damage was known and understood years ahead of time (Travis, 2005). Yet the socio-economic and political impacts of the hurricane were enormous. Likewise, the Elbe River caused severe damage before 2002 and again in April 2006 – despite comprehensive event analyses and lessons-learned processes. The failure to mitigate and respond more effectively to natural hazards – which in both cases have been predicted in theory for many years, and forecast with startling accuracy days ahead of time – raises a key question: What is blocking the transfer and implementation of existing research-based knowledge?

Results of a recent study on vulnerability assessments underscore the importance of viewing knowledge production as a systematic and emergent inquiry process, embedded in a collaborative partnership between scientists, policy makers and practitioners for the purpose of generating actionable scientific knowledge (Weichselgartner and Kasperson, 2010). The authors found a number of functional, structural and social barriers inhibiting the joint production of applicable knowledge by experts and decision makers: divergent objectives, needs, scope and priorities; different institutional settings and standards, as well as differing cultural values, understanding and mistrust. These factors and related pitfalls, such as failing to appreciate context and to connect different scales, were observed during and after Katrina and the Elbe River floods. It is assumed that the losses are ultimately due to an unsatisfactory transformation of existing knowledge – that is, converting theoretical research findings into concrete actions in practical disaster management (Weichselgartner and Obersteiner, 2002).

Both disasters provided evidence that scientific knowledge will no longer suffice, at least in those cases where the accuracy and reliability reached within the scientific community will fail to influence decision makers in policy and practice. Not only is 'reliable knowledge' (i.e. ensuring robustness through conventional discipline-bound norms) needed, but there also exists a need to be sensitive to a much wider range of social implications to produce socially robust and context-sensitive knowledge (i.e. a more locally embedded, historically and socially contingent knowledge production).

Conclusion

Both Hurricane Katrina and the Elbe River floods plainly demonstrated that highly technical societies are also vulnerable to natural hazards. Moreover, they destroyed confidence in the security of people's living conditions and in the institutions responsible for disaster protection. The disasters revealed how traditional risk management policy does not sufficiently consider natural hazards as a product of the interrelations between the physical environment, society and man-made environment. Both scientists and practitioners frequently operate with a narrow focus on geophysical processes and a preference for technical fixes and structural measures without adequately integrating social, economic, political and legislative contexts, resulting in an insufficient consideration of mitigation as a cross-cutting issue and inadequate mitigation capacities. Consequently, hazard protection is a socially isolated activity and disaster response an event-focused reaction – both highly professionalized but seldom viewed as an integral part of a larger development context.

In both Germany and the US, decision making structures in flood risk management are complex and a variety of competences govern the management of water bodies. The bureaucracy around floodplain management is convoluted and different aspects of water are managed separately – for instance, flood management, agriculture, environmental protection and recreation are handled by different organizations with different responsibilities, in different modes with conflicting objectives, and with different spatial and temporal scales. Separated responsibilities and unclear competences cause difficulties in coordinating disaster response activities and, therefore, the decentralized management systems need to improve structures of responsibility and knowledge about mechanisms, regulations and possibilities.

The federal system in Germany faces the challenge of the continued efficient use of disaster-response resources after dismantling bureaucracy and changing the administrative structures. Another obstacle to effective flood risk management is the lack of a comprehensive planning tool and legislation that includes all the various sectors and administrative levels. For many decades in German flood prevention policy, structural approaches have dominated non-structural strategies such as land-use management or hazard zoning. By learning from neighbours like Switzerland and The Netherlands, it is increasingly acknowledged that flood losses can only effectively be reduced if, in addition to technical flood protection, spatial planning regulates land use in flood prone areas and more room is given to the rivers.

The tradition of voluntary services in emergency management and a diverse response system has proven to be a solid base for disaster management in Germany. Through the tradition of voluntary work in disaster relief (almost two million people work in the voluntary fire brigades, in volunteer organizations, civilian service and the voluntary social year), a culture of solidarity and resilience has developed. The floods of 2002 revealed that insured persons had taken better precautionary measures than uninsured persons and, similarly, implemented emergency measures. This contradicts the widespread notion that an insured person is not interested in reducing damage.

Disaster management that operates in a top–down fashion and ignores the local knowledge, social dynamics and involvement of those most at risk provides a paternalistic view of 'what's best'. Even the most advanced forecasting capabilities and response capacities are limited in effectiveness when the at-risk people are not made visible in emergency planning, are not reached in the forecasting programmes or not included in decision making processes. Cross-cultural comparisons of disasters are a feasible way to gain a more locally embedded, context-sensitive, and historically and socially contingent knowledge, providing an essential step towards a collective learning from disasters.

References

BfG (2002) 'Das Augusthochwasser 2002 im Elbegebiet', Bundesanstalt für Gewässerkunde, Koblenz

Bundesministerium der Justiz (2002) 'Bund und Länder einigen sich über die Aufteilung der Mittel im Fonds Aufbauhilfe', Press Release No. 69/02, 20 December 2002, Berlin

Cutter, S. L., Emrich, C. T., Mitchell, J. T., Boruff, B. J., Gall, M., Schmidtlein, M. C., Burton, C. G. and Melton, G. (2006) 'The long road home: Race, class, and recovery from hurricane Katrina', *Environment*, vol 48, no 2, pp8–20

Demirbilek, Z. (ed) (2010) 'A forensic analysis of hurricane Katrina's impact: Methods and findings', *Ocean Engineering SI*, vol 37, no 1, pp1–154

Diamond, J. (2005) *Collapse: How Societies Choose to Fail or Survive*, Viking Penguin, New York

DKKV (2004) *Flood Risk Reduction in Germany: Lessons Learned from the 2002 Disaster in the Elbe Region*, DKKV Publication No. 29e, German Committee for Disaster Reduction, Bonn

Kates, R. W., Colten, C. E., Laska, S. and Leatherman, S. P. (2006) 'Reconstruction of New Orleans after Hurricane Katrina: A research perspective' *PNAS*, vol 103, no 40, pp14653–14660

Kirchbach, H. P. von, Franke, S., Biele, H., Minnich, L., Epple, M., Schäfer, F., Unnasch, F. and Schuster, M. (2002) 'Bericht der Unabhängigen Kommission der Sächsischen Staatsregierung Flutkatastrophe 2002', www.emergency-management.net/pdf/14_10_2003/KommissionsberichtElbeflutkatastrophe2002.pdf, accessed 9 January 2010

LfUG (2004) 'Event Analysis: August 2002 Flood in the Watercourses of the Osterzgebirge Mountains', Management Report, Saxon State Office for Environment and Geology, Dresden

Lübken, U. and Mauch, C. (2004) 'Natural disasters in transatlantic perspective: River floods in German and U.S. history', *Bulletin of the German Historical Institute*, no 35, pp99–111

Mauch, C. and Pfister, C. (eds) (2009) *Natural Disasters, Cultural Responses: Case Studies Toward a Global Environmental History*, Lexington Books, Lanham, MD

Mechler, R. and Weichselgartner, J. (2003) 'Disaster Loss Financing in Germany: The Case of the Elbe River Floods 2002', IIASA Interim Report IR-03-021, International Institute for Applied Systems Analysis, Laxenburg

Pilkington, E. (2009) 'Victims of flooding during Hurricane Katrina win compensation', *The Guardian*, 19 November, www.guardian.co.uk/world/2009/nov/19/hurricane-katrina-victims-compensation, accessed 25 December 2009

Poliwoda, G. (2007) 'Learning from disasters: Saxony fights the floods of the River Elbe 1784-1845', *Historical Social Research*, vol 32, no 3, pp169–199

Roggema, R. (2009) *Adaptation to Climate Change: A Spatial Challenge*, Springer, Heidelberg

Rosenthal, U. and 't Hart, P. (1998) *Flood Response and Crisis Management in Western Europe: A Comparative Analysis*, Springer, Heidelberg

Select Bipartisan Committee to Investigate the Preparation for and Response to Hurricane Katrina (2006) 'A Failure of Initiative: Final Report of the Select Bipartisan Committee to Investigate the Preparation for and Response to Hurricane Katrina', US Government Printing Office, Washington DC, www.gpoaccess.gov/katrinareport/fullreport.pdf, accessed 9 January 2010

SMUL (2002) 'Hochwasserschutz in Sachsen', Materialien zur Wasserwirtschaft, Sächsisches Staatsministerium für Umwelt und Landwirtschaft, Dresden

Travis, J. (2005) 'Scientists' fears come true as hurricane floods New Orleans', *Science* vol 309, pp1656–1659

Weichselgartner, J. and Bertens, J. (2002) 'Natural Disaster Reduction in Europe: A Don Quixotic Project in a Changing World?', in C. A. Brebbia (ed) *Risk Analysis III*, WIT Press, Southampton, pp233–242

Weichselgartner, J. and Obersteiner, M. (2002) 'Knowing sufficient and applying more: Challenges in hazards management', *Global Environmental Change Part B: Environmental Hazards*, vol 4, nos 2–3, pp73–77

Weichselgartner, J. and Sendzimir, J. (2004) 'Resolving the paradox: Food for thought on the wider dimensions of natural disasters', *Mountain Research and Development*, vol 24, no 1, pp4–9

Weichselgartner, J. and Kasperson, R. E. (2010) 'Barriers in the science-policy-practice interface: Toward a knowledge-action-system in global environmental change research', *Global Environmental Change*, vol 20, no 2, pp266–277

9

Social Dynamics of Unnatural Disasters: Parallels between Hurricane Katrina and the 2003 European Heat Wave

Melanie Gall

Introduction

The public health field uses a succinct term to describe the phenomenon when more deaths occur than would be expected under normal conditions in a given population: excess mortality. Excess mortality refers to any premature death or death prior to the average life expectancy for a particular demographic group (Mosby, 2009).

A disaster such as a heat wave or hurricane generates excess mortality. Hurricane Katrina, for example, caused 1833 deaths (Knabb et al, 2005) and the 2003 European heat wave triggered between 30,000 and 50,000 excess deaths[1] in central and southern Europe (Brücker, 2005). Although the origins of both events were natural, there was nothing natural with regard to the resulting deaths. Many fatalities could have been prevented through better emergency preparedness and response, and through a better understanding and targeting of vulnerable populations. In other words, if people at risk had been evacuated, warned and/or assisted in a timely fashion, suffering and deaths could have been reduced or avoided altogether. Existing research shows that vulnerable groups such as special needs populations or the elderly suffer disproportionately in times of crisis because they lack the resources or abilities to prepare for, respond to and/or recover from the event (Ngo, 2001). This chapter outlines the dimensions of vulnerability and highlights reoccurring themes through the lenses of Hurricane Katrina and the 2003 European heat wave. Despite the differences in hazard type, the similarities in terms of vulnerability, particularly social vulnerability, are striking as the following discussion will show.

Vulnerability

Broadly defined, vulnerability to environmental hazards represents the potential for loss (Mitchell, 1989) or the capacity to suffer harm and being adversely affected (Kates, 1985). A multitude of other, mostly complementary definitions exists, which reflect the diversity of methodologies and disciplines studying vulnerability (Eakin and Luers, 2006). For some, vulnerability is driven by exposure and biophysical conditions (Burton et al, 1993) such as living in a floodplain or in close proximity to chemical facilities, while others identify chronic or acute societal conditions such as limited access to political and economic power and resources as the driving forces (Wisner et al, 2003).[2]

Ultimately it is often a combination of both exposure and societal conditions (subsequently referred to as social vulnerability) that produces place-specific vulnerability (Cutter, 1996; Turner et al, 2003). As this comparison between Hurricane Katrina and the European heat wave highlights, the degree of suffering and the differential impact on society can be largely explained by social vulnerability – independent of the exposure. Since this chapter concentrates on social vulnerability, a detailed discussion of important aspects of physical vulnerability as pertaining to New Orleans (e.g. situated below sea-level, coastal location, etc.) or Paris (e.g. urban heat island effect, ozone pollution and so forth) is therefore omitted.

Social vulnerability as the culprit of misery

Social vulnerability spans a multidimensional space that incorporates social, political, institutional and economic conditions unique to specific individuals, an organization, community or a place at a specific point in time (Bohle et al, 1994). It may even include conditions of the built environment (Cutter and Finch, 2008) and environmental or ecological components particularly where issues of resilience and sustainability (Adger et al, 2005) are concerned or within the context of developing countries (Adger, 1999).

Widely accepted factors of social vulnerability – that is, conditions that reduce the ability to cope with the impacts of a hazardous event – are age, gender, socio-economic status, race/ethnicity, critical infrastructure, social networks, health status and governance (Fothergill, 1996; Fothergill et al, 1999; Cutter et al, 2003; Lindsay, 2003; Wisner et al, 2003; Fothergill and Peek, 2004; Phillips and Morrow, 2007). Since most social vulnerability assessments require the consideration of context- and place-specific circumstances such as those addressed by Weichselgartner and Breviere (Chapter 8), it is almost impossible to define a globally applicable set of standard factors that influence social vulnerability (Birkmann, 2005). Despite the lack of a consensus on the key determinants of social vulnerability, it is possible to draw comparisons and find similarities in social vulnerability between vastly different places and/or societies. This chapter does just that and explores recurring themes of social vulnerability as experienced by the residents of New Orleans in 2005 and Parisians in 2003.

Different hazards but similar outcomes

The physical exposure to a heat wave versus a hurricane is as different as the impact and damage suffered from either event. A hurricane or tropical cyclone is a meteorological event – a low pressure system with organized circulation that originated in tropical waters and, depending on intensity, can bring with it heavy rainfall, tornadoes, flooding and/or storm surge. Due to their life cycle and development stages, hurricanes are easy to detect and track by meteorological satellites. In the United States, the National Weather Service operates an advanced network of detection, tracking and forecasting tools that allow emergency managers and public officials to prepare areas at risk and take the necessary steps to evacuate people, preposition resources and secure properties – days prior to the anticipated landfall along the Mississippi–Louisiana state border. Defying all mitigation and preparedness efforts in August of 2005, Hurricane Katrina caused more than $80 billion in direct property damage and more than 1500 fatalities (Blake et al, 2007) with drowning being the leading cause of death (Bourque et al, 2006).

Heat waves are meteorological events as well but detecting, forecasting or even tracking them along with their impacts is challenging. A heat wave in a cooler climate might be identical to normal weather conditions in a warmer climate. Consequently, there is no universal definition of what constitutes a heat wave since it varies from place to place. The American National Weather Service defines a heat wave as 'a period of abnormally and uncomfortably hot and unusually humid weather' whereas the French Weather Services determines a heat wave by the daily maximum temperature exceeding 30°C (Poumadère et al, 2005). Unlike hurricanes, the severity of heat waves and their threat to vulnerable populations tends to be determined based on outcomes (e.g. health-related incidents and mortality) and not necessarily climatological conditions. Responding to heat waves is therefore much more reactive than proactive in comparison to hurricanes.

In the summer of 2003, Europe faced its hottest summer since AD 1500 (Luterbacher et al, 2004). It is estimated that the record-setting heat wave caused between 30,000 and 50,000 excess deaths in Europe (Brücker, 2005) and affected Portugal, Spain, Italy, France, the United Kingdom, Germany, The Netherlands, Belgium, Switzerland and others, with France being the hardest hit (Kosatsky, 2005) (see Figure 9.1).

This chapter focuses on heat wave events as they unfolded in France and its urban areas between 1 and 20 August 2003. This time period serves as the basis for most public health studies in France given that on nine consecutive days the mean maximum temperature exceeded the seasonal norm by 11 to 12°C. Typical summer temperatures for this time of the year fluctuate around 25°C (77°F) but in 2003, August temperatures soared to 37°C (99°F) for more

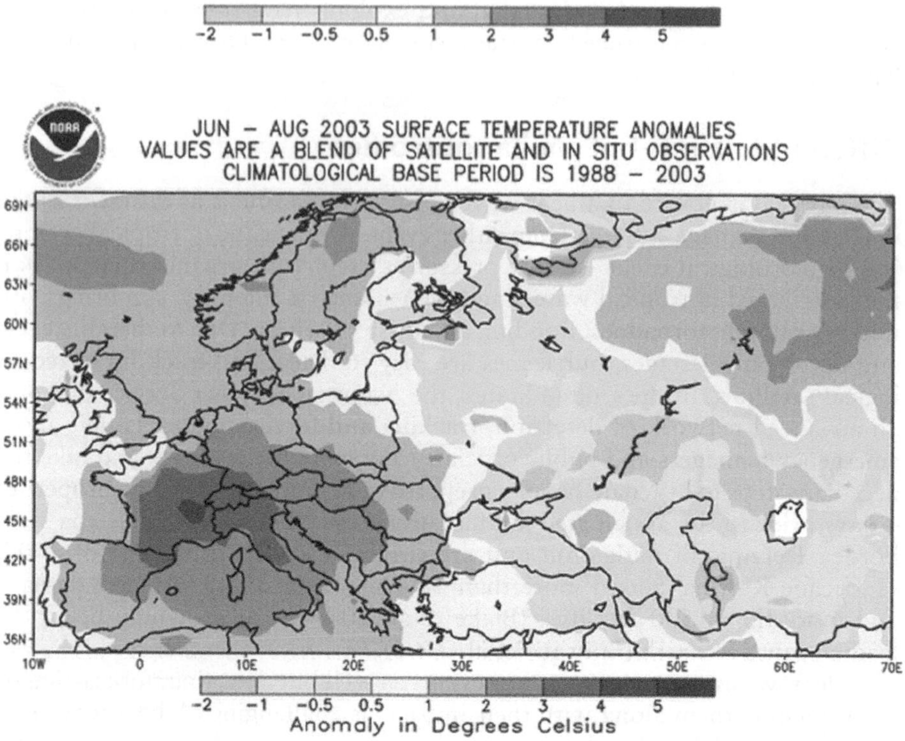

Source: NASA Earth Observatory, available from http://earthobservatory.nasa.gov/IOTD/view.php?id=3714

Figure 9.1 *Temperature differences between July 2001 and July 2003*

than a week (Fouillet et al, 2006). About 60 million people were affected and 14,800 excess deaths occurred in France alone (Pirard et al, 2005). The most frequent causes of heat-related deaths were hyperthermia, heatstroke, dehydration, cardiovascular diseases, among others (Kovats and Hajat, 2008). It should be noted that Europe-wide the unusually hot weather patterns existed well before 1 August and persisted beyond 20 August.

At first glance, drowning in hurricane waters has nothing in common with dying from dehydration. That is true until the search for commonalities moves from *how* someone died to *who* died. In both instances, it was elderly, fragile and/or marginalized individuals – that is, socially vulnerable groups. As evidenced by Henderson's study of evacuees (Chapter 5), and by Barrios' (Chapter 6) and Allen and Maret's (Chapter 7) studies of neighbourhood recovery efforts, social vulnerability at the individual level combined with social vulnerability at the community and organizational level exacerbated the devastating impacts of Katrina. This chapter shows how these vulnerabilities exacerbated the impacts of both events causing Hurricane Katrina and the 2003 European heat wave to rank among the deadliest events in their respective regions.

Social vulnerability: Individual level

The most obvious overlap between both events is the demographic of their victims. As mentioned previously, age and health status are key drivers of a person's level of social vulnerability. For the State of Louisiana, the state's Department of Health and Hospitals (2006) reported 1464 Katrina victims, of which 346 persons died outside of the state. Jonkman et al (2009) analysed information of more than 800 victims who perished in New Orleans and found that more than 50 per cent of Hurricane Katrina's victims were older than 75 years and about 85 per cent of all victims were older than 51 years (Table 9.1). Jonkman et al did not detect a gender or ethnicity/race bias and found the gender and racial make-up of New Orleans reflected in the demographics of the victims.

Table 9.1 *Fatalities breakdown by age, gender and other characteristics*

	Hurricane Katrina New Orleans[1]	*Heat wave Paris*[2]
Age		
Older than 65 years	60%	40%*
Older than 75 years	50%	70%*[3]
Gender		
Female	49%	70%*
Male	15%	40%*
Race/Ethnicity		
African-American	55%	n/a
White	40%	n/a
Hispanic	2%	n/a
Other	3%	n/a
Location		
Residence	54%	35%
Medical facilities	20%	42%
Nursing homes	10%	19%
Other	16%	4%
Total number of fatalities	853	1854[3]

*Indicates excess death percentages and NOT percentage of total number of fatalities. Excess death percentages are relative to a prior point in time; that is, excess death percentages reflect how many more (e.g. elderly) died during the summer of 2003 than in prior years. For Paris, the reference time period is mortality rates by age and sex from 1 to 19 August averaged across four consecutive years (1999–2002). While excess death calculations include ALL causes of death, the 'excess' effect is directly attributed to the heat wave.

Source: 1 Jonkman et al (2009); 2 Dhainaut et al (2004); 3 Pirard et al (2005)

The 2003 European heat wave manifested itself in a similar pattern leading to 14,800 excess deaths in France between 1 and 20 August. Paris alone accounted for 1854 excess deaths with many more victims in France's large cities such as Marseille (571 fatalities), Lyon (447 fatalities), Nice (341 fatalities), Bordeaux (318 fatalities), Toulouse (315 fatalities) and so forth (Pirard et al, 2005). The capital of France, Paris, was by far the hardest hit city with an excess mortality rate of 142 per cent; that is, the heat wave caused an increase of 142 per cent in fatalities compared to a 'normal' year. The city's elderly population was in the cross-hairs of the heat wave with excess deaths of 70 per cent for persons older than 75 years. In contrast to Hurricane Katrina, there was a pronounced adverse effect on female mortality with a 70 per cent increase in excess mortality versus 40 per cent in men (Dhainaut et al, 2004). On a technical side note, the reference time period (or normal) for calculating excess deaths was 1 to 19 August for four consecutive years (1999–2002).

The hazard burden on elderly populations as exemplified by Hurricane Katrina and the 2003 heat wave is nothing new. The elderly are highly vulnerable to natural hazards not necessarily because of their age but because they often exhibit additional conditions that increase their social vulnerability such as chronic diseases, reduced cognitive and physical abilities (special needs), low socio-economic status or reliance on fixed social security income, living alone and/or in isolated conditions (Bell et al, 1978; Ngo, 2001; Klinenberg, 2002). Furthermore, seniors often exhibit reluctance towards evacuation and other preparedness measures as well as accessing services because they underestimate the risk and threat to life and property (Phillips and Morrow, 2007).

This was also true for many victims of the heat wave and Hurricane Katrina. The above described factors specific to the social vulnerability of the elderly also explain the spatial distribution of fatalities within the city of New Orleans as well as in Paris. As Jonkman et al (2009) found, more than half of the elderly victims died at home (Table 9.1) because they could or would not evacuate – a result of the absence of evacuation assistance or lack of other means of transportation, fear of looting, physical and/or health handicaps that made evacuation or staying in shelters seem daunting, lack of financial resources and a disregard and disbelief of the seriousness of the threat. Prior to Katrina, many elderly New Orleans residents had stayed and survived other storms such as Hurricane Betsy (Laska and Morrow, 2006; Bytheway, 2007). Similarly in Paris, 92 per cent of city dwellers that died at home lived alone, 70 per cent had chronic health problems such as cardiovascular diseases, and about one fourth had no family or social ties (Poumadère et al, 2005).

Ultimately, emergency managers in New Orleans and Paris ignored the fatal threat to the most vulnerable of their citizens. They neither harnessed the long-established knowledge about the social vulnerability of elderly people nor translated it into preparedness measures and warning messages tailored and targeted at the elderly. The burden on the elderly could have been significantly reduced if their needs and concerns had been considered and addressed before either event.

Social vulnerability: Community level

The needs of the elderly as well as other highly vulnerable groups (e.g. disabled, children, poor, etc.) go beyond physical well-being and financial support. The fact that many victims of Hurricane Katrina and the 2003 heat wave died at home points again to inadequate housing conditions and insufficient hazard protection from either hazard. Poor housing conditions often go hand in hand with limited financial resource (Bolin and Stanford, 1998) but they can also be a product of local housing traditions, land-use policies, risk culture and more (Heinz Center, 2000).

In Paris, more than half of the victims lived on the two highest floors of buildings (Poumadère et al, 2005). This is of great significance given that (a) air conditioning is not a common amenity in Paris, (b) many elderly occupy '*mansardes*', which are traditional, small, often one-room apartments directly under the roof, and (c) metal roofs which are prevalent in Paris accelerate heat build-up and contribute to the urban heat island effect (Poumadère et al, 2005; Haines et al, 2006). Thus, the Parisian built environment is extremely susceptible to the adverse impacts of heat waves and instead of providing protection against the elements, it enhances the detrimental effects of heat waves.

The urban culture and way of living tends to also increase social vulnerability. The absence of social networks and the decline of family relationships, particularly in Western cultures, diminish the value of inter-personal 'alert and monitoring' systems. In Paris, victims that died at home alone had no family or social ties (about 25 per cent), only had contact with neighbours or friends (25 per cent) or only had interactions with medical and social services (less than 20 per cent). Out of all home victims, only 30 per cent were in regular contact with family (Poumadère et al, 2005).

In order to mitigate adverse health effects from heat, residents need to be aware of them. This requires educational campaigns to explain the signs of dehydration, heat exhaustion, heat cramps and so on, and increase the awareness of effective measures to curb adverse heat effects, for example, ventilation, spend time in air-conditioned buildings and so forth (Kovats and Hajat, 2008; Ebi, 2009). For instance, the city of Marseille incurred excess mortality of 25 per cent – much lower rates than Paris. Some attribute this to Marseille's experience with heat waves in 1983 and 2003, which seemingly left the city and its residents with a heightened awareness and with increased coping capacity with regard to the 2003 event (Vandentorren et al, 2004).

Coping capacity is a corner stone of social vulnerability (Yohe and Tol, 2002): the lower the coping capacity, the higher the social vulnerability. The coping capacity of New Orleans to manage the onslaught of storm waters into its flood protection system was substandard to say the least. A vicious cycle of reliance on structural flood mitigation and the expectation of residents to be – or get – protected combined with a crumbling flood protection system provided the basis for Katrina's deadly punch. In the past, residents of New Orleans were not actively required to mitigate flood events. In fact, through failed land-use policies the city put people (and nursing homes and hospitals, etc.) in harm's way, while misguided federal flood insurance policies relieved residents from insurance requirements such as elevating homes or purchasing flood insurance (Burby, 2006). The city of New Orleans was anything but flood-proof nor were its residents flood- or storm-ready – just as Parisians were not prepared or experienced in realizing and counteracting the effects of the record-breaking heat wave.

This is where governmental organizations and entities need to step in to help vulnerable groups to prepare, respond and recover. However, the perpetuation of business as usual by governmental entities amplified social vulnerability at the individual and community level in Paris as well as in New Orleans. So, where did the governmental and emergency management systems of both cities fail its vulnerable populations?

Social vulnerability: Organizational level

Local governments in New Orleans and Paris ignored the seriousness of the respective situations and wasted precious time for preparedness and response actions. In France, no national policy existed on how to prepare for or respond to heat waves. In the initial phase of the heat wave, emergency medical services and funeral homes were the first confronted with the impacts, and thus the first to realize that the summer of 2003 was different from previous years. Their alerts to public officials, however, could not trigger an appropriate or effective government response. In fact, the French Prime Minister and the Minister of Health downplayed the graveness of the situation, restricted the publication of death figures, and dismissed media reports as sensationalism (Lagadec, 2004; Poumadère et al, 2005). This was a dramatic mistake considering that raising awareness (e.g. heat alerts), educating the public (e.g. on health risk, encouraging neighbourly concern) and providing preventive medical advice (e.g. drinking more fluids, staying in air-conditioned areas) are key factors in reducing heat-related mortality (Basu and Samet, 2002).

A study on nuclear risk perception between France and the United States revealed high levels of trust and reliance of French citizens in their government and particularly public health officials. According to Slovic et al (2000), the French are more fatalistic compared to US residents in the sense that they tend to feel health decisions should be left to public health experts because they trust officials and feel confident that public health officials will take the necessary steps to protect the public. (See Dowty et al (Chapter 2) for an overview of the effects of a fatalistic organizational culture in a United States government agency.) This seems surprising given that some researchers attest to the French elites' secretive and reluctant attitudes (Porter, 1996; Lagadec, 2004). One can conclude that the reliance on good governance by the French public combined with stonewalling and a 'Deny, Deflect, Defend' approach (Klinenberg, 2002, p168) by public health officials caused significant delays in emergency response, for which many French citizens paid with their lives.

Even worse, instead of accepting the need for action and ordering a surge in hospital capacity and other medical services, French cities lived through the heat wave with fewer rather than more resources. The month of August is traditionally a time when many residents are on vacation – meaning that many nurses, doctors, government officials, and so on were on leave. As a result, governmental as well as medical services were understaffed and could not handle the unfolding crisis (Lagadec, 2004).

In New Orleans, understaffing and a complete breakdown of emergency services was a contributing factor as well (among many others). Equally, the 'Deny, Deflect, Defend' approach was the initial reaction of local, state and federal officials, although live media coverage from New Orleans documented the true situation and the lack of action. Many attribute failures at local, state and federal levels to the post-9/11 changes in the US emergency management system, which resulted in unfamiliar chains of command, protocols and policies (Bea, 2005; Tierney, 2006b; Gall and Cutter, 2007). Additionally, the rampant political corruption and cronyism in New Orleans provided a facilitating environment and severely eroded the organizational capacities of the city (Brinkley, 2006).

Instead of reiterating these dramatic organizational failures in more detail, it is important to point out that New Orleans and thus local governmental entities were seemingly either unaware or simply ignored the high levels of social vulnerability present at the individual and community level. It appears that aside from the fictitious 2004 Hurricane Pam exercise, local agencies did not consider socially vulnerable groups in their planning efforts. As such, the city was not ready nor did it imagine the challenges that the many impoverished, carless, disabled, impaired or pet-owning residents would face when less than 24 hours before landfall, Mayor Ray Nagin called for the city's first ever mandatory (ad hoc) evacuation (Fussel, 2006). There was neither time to stockpile or equip shelters of last resort such as the Superdome with sufficient supplies to accommodate the influx of evacuees, nor time for hospitals, for example, to adapt to the new situation.

Jonkman et al (2009) found about 20 per cent of fatalities occurred at hospitals and shelters, due to lack of medical services and the disruption of power and water services (see Table 9.1). These deaths could have been avoided by timely evacuations, proper sheltering and mass care, pre-positioning of generators, fuel, water, food, medical supplies and so forth. New Orleans was warned days ahead of Katrina's landfall and had theoretically sufficient time to get ready, but organizational inefficacy aggravated individual and community vulnerabilities and made a horrific situation worse.

The burden of unnatural disasters

Riding out a storm in a city as physically vulnerable as New Orleans, in homes with rarely any flood protection (including insurance), with local and state governments caught unprepared, by residents unaware or underestimating the risks, was a recipe for an unnatural disaster. It was the product of social vulnerabilities at the individual, community and organizational level. Likewise, having a detached governmental risk perception while unaware, unprepared and uncared-for elderly people lived through an unprecedented heat wave, in non-air-conditioned, badly ventilated upper level apartments could have only meant one thing – unnecessary deaths. Burby (2006) calls this ignorance of risk by local governments and their failure to protect residents the 'local government paradox' while Tierney (2006a) attributes the unequal burden of a disaster to the 'Matthew Effect'. The Matthew Effect causes socially vulnerable groups to suffer disproportionately while affluent populations remain fairly unscathed.

To make matters worse, there seems to be a vulnerability paradox. Plenty of empirical research and knowledge exists on the drivers of social vulnerability and their manifestation during a disaster event. One could point again to the seminal work of Klinenberg (2002) but seemingly this knowledge cannot penetrate or infiltrate local, state and federal mindsets. Issues of social vulnerability tend to find their way into planning guidelines or bills such as the Post-Katrina Emergency Management Reform Act after the fact – that is, when lives have been lost due to a lack of understanding and a lack of imagination. For instance, within a year after the 2003 disaster, France implemented the National Heat Health Watch Warning System (Système d'Alerte Canicule Santé, SACS), which was effectively tested during the 2006 heat wave (Fouillet et al, 2008) and the State of Louisiana purchased the (not yet fully functioning) evacuation tracking system 'Phoenix' to improve its embarrassing 2005 performance (Louisiana Department of Social Services, 2009).

It is seemingly impossible for local government to adequately anticipate or acknowledge the needs of their vulnerable populations during emergencies. Heat waves and hurricanes are not infrequent. In post-industrial nations, heat waves tend to kill more people than any other type of natural hazards (Borden and Cutter, 2008), and hurricanes or tropical storms affect the US Gulf coast every three to four years (Keim et al, 2007). However, learning from the past *and* anticipating future vulnerabilities seems to remain a challenge. For instance, the ongoing debate over climate change leaves little hope that preparing for more frequent and potentially more severe heat waves, hurricanes, floods and so forth is on the planning fast track in Europe or elsewhere.

In fact, emergency planners tend to plan for the status quo. But preparing for a heat wave or a hurricane as they are experienced today means preparing for yesterday. Emergency managers need to look ahead and anticipate the next big event and its unequal burden on vulnerable populations. What if a longer lasting, more intense heat wave with higher ozone levels than experienced during the summer of 2003 increases child mortality, or what if New Orleans faced a direct hit while many migrant construction workers still live in the city? Today's vulnerable populations might not necessarily be tomorrow's vulnerable populations. It is therefore necessary to incorporate lessons from the past and use imagination and empirical research to adapt to a changing climate, hazard and impacts.

Notes

1 Estimates for the 2003 heat wave appear often vastly different and depend largely on the time frame under consideration.
2 For more in-depth discussions on conceptual issues surrounding vulnerability and related constructs such as adaptive capacity, adaptation and resilience see elsewhere (e.g. Birkmann, 2006; Eakin and Luers, 2006; Gallopín, 2006; Füssel, 2007).

References

Adger, W. N. (1999) 'Social vulnerability to climate change and extremes in coastal Vietnam', *World Development*, vol 27, no 2, pp249–269

Adger, W. N., Hughes, T. P., Folke, C., Carpenter, S. R. and Rockström, J. (2005) 'Social-ecological resilience to coastal disasters', *Science*, vol 309, no 5737, pp1036–1039

Basu, R. and Samet, J. M. (2002) 'Relation between elevated ambient temperature and mortality: A review of the epidemiologic evidence', *Epidemiologic Reviews*, vol 24, no 2, pp190–202

Bea, K. (2005) *The National Preparedness System: Issues in the 109th Congress*, Congressional Research Service, Washington DC

Bell, B. D., Kara, G. and Batterson, C. (1978) 'Service utilization and adjustment patterns of elderly tornado victims in an American disaster', *Mass Emergencies*, vol 3, nos 2–3, pp71–81

Birkmann, J. (2005) 'Expert Workshop: Measuring Vulnerability', United Nations University, Institute for Environment and Human Security (UNU-EHS), Bonn, Germany

Birkmann, J. (ed) (2006) *Measuring Vulnerability to Natural Hazards: Towards Disaster Resilient Societies*, United Nations University Press, New York

Blake, E. S., Rappaport, E. N. and Landsea, C. W. (2007) *The Deadliest, Costliest, and Most Intense United States Tropical Cyclones from 1851 to 2006 (and other frequently requested hurricane facts)*, National Hurricane Center (NHC), Miami, FL

Bohle, H. G., Downing, T. E. and Watts, M. J. (1994) 'Climate change and social vulnerability: The sociology and geography of food insecurity', *Global Environmental Change*, vol 4, pp37–48

Bolin, R. and Stanford, L. (1998) 'The Northridge earthquake: Community-based approaches to unmet recovery needs', *Disasters*, vol 22, no 1, pp21–38

Borden, K. A. and Cutter, S. L. (2008) 'Spatial patterns of natural hazards mortality in the United States', *International Journal of Health Geographics*, vol 7, article id 64, http://www.ij-healthgeographics.com/content/7/1/64, accessed 22 December 2010

Bourque, L. B., Siegel, J. M., Kano, M. and Wood, M. M. (2006) 'Weathering the storm: The impact of hurricanes on physical and mental health', *Annals of the American Academy of Political and Social Science*, vol 604, no 1, pp129–151

Brinkley, D. (2006) *The Great Deluge: Hurricane Katrina, New Orleans, and the Mississippi Gulf Coast*, William Morrow, New York

Brücker, G. (2005) 'Vulnerable populations: Lessons learnt from the summer 2003 heat waves in Europe' (editorial), *Eurosurveillance*, vol 10, no 7, article id 551, www.eurosurveillance.org/ViewArticle.aspx?ArticleId=551, accessed 27 March 2010

Burby, R. J. (2006) 'Hurricane Katrina and the paradoxes of government disaster policy: Bringing about wise governmental decisions for hazardous areas', *Annals of the American Academy of Political and Social Science*, vol 604, no 1, pp288–332

Burton, I., Kates, R. W. and White, G. F. (1993) *The Environment as Hazard*, 2nd edn, Guildford Press, New York

Bytheway, B. (2007) 'The evacuation of older people: The case of Hurricane Katrina', *Understanding Katrina: Perspectives from the Social Sciences*, understandingkatrina.ssrc.org/Bytheway/, accessed 29 March 2010

Cutter, S. L. (1996) 'Vulnerability to environmental hazards', *Progress in Human Geography*, vol 20, no 4, pp529–539

Cutter, S. L. and Finch, C. (2008) 'Temporal and spatial changes in social vulnerability to natural hazards', *Proceedings of the National Academy of Sciences of the United States of America*, vol 105, no 7, pp2301–2306

Cutter, S. L., Boruff, B. J. and Shirley, W. L. (2003) 'Social vulnerability to environmental hazards', *Social Science Quarterly*, vol 84, no 1, pp242–261

Dhainaut, J. F., Claessens, Y. E., Ginsburg, C. and Riou, B. (2004) 'Unprecedented heat-related deaths during the 2003 heat wave in Paris: Consequences on emergency departments', *Critical Care*, vol 8, no 1, pp1–2

Eakin, H. and Luers, A. L. (2006) 'Assessing the vulnerability of social-environmental systems', *Annual Review of Environment and Resources*, vol 31, pp365–394

Ebi, K. L. (2009) 'Facilitating climate justice through community-based adaptation in the health sector', *Environmental Justice*, vol 2, no 4, pp191–195

Fothergill, A. (1996) 'Gender, risk, and disaster', *International Journal of Mass Emergencies and Disasters*, vol 14, no 1, pp33–56

Fothergill, A. and Peek, L. A. (2004) 'Poverty and disasters in the United States: A review of recent sociological findings', *Natural Hazards*, vol 32, no 1, pp89–110

Fothergill, A., Maestas, E. G. M. and Darlington, J. D. (1999) 'Race, ethnicity and disasters in the United States: A review of the literature', *Disasters*, vol 23, no 2, pp156–173

Fouillet, A., Rey, G., Laurent, F., Pavillon, G., Bellec, S., Guihenneuc-Jouyaux, C., Clavel, J., Jougla, E. and Hemon, D. (2006) 'Excess mortality related to the August 2003 heat wave in France', *International Archives of Occupational and Environmental Health*, vol 80, no 1, pp16–24

Fouillet, A., Rey, G., Wagner, V., Laaidi, K., Empereur-Bissonnet, P., Le Tertre, A., Frayssinet, P., Bessemoulin, P., Laurent, F., De Crouy-Chanel, P., Jougla, E. and Hemon, D. (2008) 'Has the impact of heat waves on mortality changed in France since the European heat wave of summer 2003? A study of the 2006 heat wave', *International Journal of Epidemiology*, vol 37, no 2, pp309–317

Fussel, E. (2006) 'Leaving New Orleans: Social stratification, networks, and hurricane evacuation', *Understanding Katrina: Perspectives from the Social Sciences*, understandingkatrina.ssrc.org/Fussell/, accessed 5 April 2010

Füssel, H. M. (2007) 'Vulnerability: A generally applicable conceptual framework for climate change research', *Global Environmental Change*, vol 17, no 2, pp155–167

Gall, M. and Cutter, S. L. (2007) '2005 Events and Outcomes: Hurricane Katrina and Beyond', in C. B. Rubin (ed) *Emergency Management: The American Experience 1900–2005*, The Public Entity Risk Institute (PERI), Fairfax, VA

Gallopín, G. C. (2006) 'Linkages between vulnerability, resilience, and adaptive capacity', *Global Environmental Change*, vol 16, no 3, pp293–303

Haines, A., Kovats, R. S., Campbell-Lendrum, D. and Corvalan, C. (2006) 'Climate change and human health: Impacts, vulnerability, and mitigation', *Lancet*, vol 367, no 9528, pp2101–2109

Heinz Center (2000) *The Hidden Costs of Coastal Hazards: Implications for Risk Assessment and Mitigation*, Island Press, Covello, CA

Jonkman, S. N., Masskant, B., Boyd, E. and Levitan, M. L. (2009) 'Loss of life caused by the flooding of New Orleans after Hurricane Katrina: Analysis of the relationship between flood characteristics and mortality', *Risk Analysis*, vol 29, no 5, pp676–698

Kates, R. W. (1985) 'The Interaction of Climate and Society', in R. W. Kates, J. H. Ausubel and M. Berberian (eds) *Climate Impact Assessment, SCOPE 27*, Wiley, New York

Keim, B., Muller, R. A. and Stone, G. W. (2007) 'Spatiotemporal patterns and return periods of tropical storms and hurricane strikes from Texas to Maine', *Journal of Climate*, vol 20, no 14, pp3498–3509

Klinenberg, E. (2002) *Heat wave – A social autopsy of disaster in Chicago*, University of Chicago Press, Chicago

Knabb, R. D., Rhome, J. R. and Brown, D. P. (2005) 'Tropical Cyclone Report: Hurricane Katrina, 23–30 August 2005', National Hurricane Center, Miami, FL

Kosatsky, T. (2005) 'The 2003 European heat waves' (editorial), *Eurosurveillance*, vol 10, no 7, article id 552, www.eurosurveillance.org/ViewArticle.aspx?ArticleId=552, accessed March 27, 2010

Kovats, R. S. and Hajat, S. (2008) 'Heat stress and public health: A critical review', *Annual Review of Public Health*, vol 29, pp41–55

Lagadec, P. (2004) 'Understanding the French 2003 heat wave experience: Beyond the heat, a multi-layered challenge', *Journal of Contingencies and Crisis Management*, vol 12, no 4, pp160–169

Laska, S. and Morrow, B. H. (2006) 'Social vulnerabilities and Hurricane Katrina: An unnatural disaster in New Orleans', *Marine Technology Society Journal*, vol 40, no 4, pp16–26

Lindsay, J. R. (2003) 'The determinants of disaster vulnerability: Achieving sustainable mitigation through population health', *Natural Hazards*, vol 28, pp291–304

Louisiana Department of Health and Hospitals (2006) 'Reports of Missing and Deceased', www.dhh.louisiana.gov/offices/page.asp?ID=192&Detail=5248, accessed 8 March 2010

Louisiana Department of Social Services (2009) 'Evacuee Registration Plan', Louisiana Department of Social Services, Baton Rouge, LA

Luterbacher, J., Dietrich, D., Xoplaki, E., Grosjean, M. and Wanner, H. (2004) 'European seasonal and annual temperature variability, trends, and extremes since 1500', *Science*, vol 303, no 5663, pp1499–1503

Mitchell, J. K. (1989) 'Hazards Research', in G. L. Gaile and C. J. Willmott (eds) *Geography in America*, Merrill, Columbus, OH

Mosby (2009) *Mosby's Medical Dictionary*, 8th edn, Elsevier

National Weather Service (2009) 'National Weather Service Glossary', www.weather.gov/glossary/, accessed 27 March 2010

Ngo, E. B. (2001) 'When disasters and age collide: Reviewing vulnerability of the elderly', *Natural Hazards Review*, vol 2, no 2, pp80–89

Phillips, B. and Morrow, B. H. (2007) 'Social science research needs: Focus on vulnerable populations, forecasting, and warnings', *Natural Hazards Review*, vol 8, no 3, pp61–68

Pirard, P., Vandentorren, S., Pascal, M., Laaidi, K., Le Tertre, A., Cassadou, S. and Ledrans, M. (2005) 'Summary of the mortality impact assessment of the 2003 heat wave in France', *Eurosurveillance*, vol 10, no 7, article id 554

Porter, T. M. (1996) *Trust in Numbers: The Pursuits of Objectivity in Science and Public Life*, Princeton University Press, Princeton, NJ

Poumadère, M., Mays, C., Le Mer, S. and Blong, R. (2005) 'The 2003 heat wave in France: Dangerous climate change here and now', *Risk Analysis*, vol 25, no 6, pp1483–1494

Slovic, P., Flynn, J., Mertz, C. K., Poumadère, M. and Mays, C. (2000) 'Nuclear Power and the Public: A Comparative Study of Risk Perception in France and the United States', in O. Renn and B. Rohrmann (eds) *Cross-cultural Risk Perception: A Survey of Empirical Studies*, Kluwer, Amsterdam

Tierney, K. J. (2006a) 'Foreshadowing Katrina: Recent sociological contributions to vulnerability science', *Contemporary Sociology*, vol 35, no 3, pp207–212

Tierney, K. J. (2006b) 'The red pill', *Understanding Katrina: Perspectives From the Social Sciences*, understandingkatrina.ssrc.org/Tierney/, accessed 5 April 2010

Turner, B. L. I., Kasperson, R. E., Matson, P. A., McCarthy, J. J., Corell, R. W., Christensen, L., Eckley, N., Kasperson, J. X., Luers, A., Martello, M. L., Polsky, C., Pulsipher, A. and Schiller, A. (2003) 'A framework for vulnerability analysis in sustainability science', *PNAS (Proceedings of the National Academy of Sciences of the United States of America)*, vol 100, no 14, pp8074–8079

Vandentorren, S., Suzan, F., Medina, S., Pascal, M., Maulpoix, A., Cohen, J. C. and Ledrans, M. (2004) 'Mortality in 13 French cities during the August 2003 heat wave', *Am J Public Health*, vol 94, no 9, pp1518–1520

Wisner, B., Blaikie, P., Cannon, T. and Davis, I. (2003) *At Risk: Natural Hazards, People's Vulnerability, and Disasters*, 2nd edn, Routledge, London

Yohe, G. and Tol, R. S. J. (2002) 'Indicators for social and economic coping capacity: Moving toward a working definition of adaptive capacity', *Global Environmental Change*, vol 12, pp25–40

10
After Disasters: Emergences of National Insecurity in Sri Lanka

Vivian Y. Choi

> *I will not say that the disaster is absolute; on the contrary, it disorients the absolute. It comes and goes, errant disarray, and yet with the imperceptible but intense suddenness of the outside as an irresistible or unforeseen resolve which would come to us from beyond the confines of decision.*
>
> Maurice Blanchot (1986)

The disaster is constantly in motion

In *Sorting Things Out: Classification and Its Consequences* (1999), Geoffrey C. Bowker and Susan Leigh Star detail the history of tuberculosis to suggest that its classification as a disease is a complex process, and, moreover, that it is articulated contingently across time, spaces and human bodies. In my research on the 2004 Indian Ocean tsunami in Sri Lanka, I see 'disaster' as a category similarly: its impacts and effects vary and move across people and places, and accordingly, are shaped and mediated differently. Like my interlocutors in this edited volume, my research is vested in the multifaceted and multi-scalar ways in which disaster spirals and morphs, even years after its supposed end. In Sri Lanka, as in New Orleans after Hurricane Katrina, when the dark and dangerous waters of the tsunami receded, the disaster laid bare some of the most embedded of social issues. Indeed, as Anthony Oliver-Smith asserts, disasters are notoriously difficult to study because they are multidimensional and processual (Oliver-Smith, 2002). As occurrences that can sweep across every aspect of human life – impacting environments and social, political and biological conditions – accounting for and encompassing this multidimensionality pose challenges for any researcher. Disaster, thus, understood as a process that unfolds through time and space, requires the manager and researcher of disaster to account for its many shifting faces and articulations (Fortun, 2001; Oliver-Smith, 2002).

As other chapters in this volume illustrate, in the maelstrom that is the aftermath of disaster, practices and attempts to order and manage disaster come to the fore, calling for a sustained and critical examination of these processes. For example, as Frickel and Vincent (Chapter 1) attest to, the Environmental Protection Agency (EPA) responded immediately with environmental testing of New Orleans sediment and soil, producing knowledge, and more significantly, 'knowledge gaps' regarding soil contamination and chemical risk. Certain knowledge gaps that emerge in their assessment of EPA reports illustrate the politics and institutional logic of efficiency through which information about environmental risk in the aftermath of Katrina. Dowty et al (Chapter 2) utilize an organizational culture lens to understand institutional responses and their perceived failures, while Bowden (Chapter 3) examines Katrina as a 'system accident', or the unanticipated interactions of multiple failures. In the case of Hurricane Katrina in New Orleans, Bowden turns to a socio-ecological perspective to better understand how social issues, ecological issues and technical issues all coincided to create an unusual level of social disorder.

After the tsunami in Sri Lanka, the situation was no different: The Indian Ocean tsunami in December 2004, resulting from a massive earthquake in the notorious 'Ring of Fire' off the western coast of northern Sumatra, Indonesia, caused unprecedented death and damage. Sri Lanka, after Indonesia, was the most devastated, with an estimated 36,000 deaths and over a million people displaced.

In this chapter I discuss the ways in which the management of disaster and future disasters in Sri Lanka has unfolded and continues to unfold. In particular, I give special attention to the ways in which the management of disaster in Sri Lanka articulates with the political and social landscape of war. To do so, I also examine management practices and the category of disaster – both 'natural' and 'man-made' – through examination of the relationship between so-called 'nature and culture'. Using this lens, nature is thus made in the same way culture is often perceived and the two constantly mediate each other, in a co-constitutive relationship, or what has been termed 'hybridization'. The study of social consequences and relationships that emerge from hybridizations have long been a hallmark of the social studies of science (Callon, 1986; Latour, 1993; Law, 2004).

Dissolving the boundary between nature and culture has been fruitful for disaster researchers and practitioners who largely accept that natural disasters are indeed, 'not natural'. Much recent scholarship on disasters has revealed as much (for example: Quarentelli, 1985; Oliver-Smith, 2004; Dawdy, 2006; Masqualier, 2006; Smith, 2006). Moreover, disasters have the potential to highlight and even exacerbate existing social tensions, at the same time also creating new ones. While much of the research and literature fixates on these issues in terms of 'social vulnerability' (Wisner, 1993; Hewitt, 1997; and for an interesting discussion on 'vulnerability' as a product of 'Western discourse' see Bankoff, 2001), I turn to the emerging governmental and management logics in Sri Lanka after the Indian Ocean tsunami that attempt to divert and manage future disasters, both natural and man-made. The logic that undergirds these new systems of governance has also internalized the 'un-naturalness' of disasters, and, accordingly, treats natural disasters and so-called acts of terrorism as seemingly inevitable, uncontrollable events that need to be managed. In Sri Lanka, the logic of managing disasters *and* conflict and terrorism must be understood together. There, the war is not merely the social context in which the tsunami played out, but rather the tsunami engendered a certain way of managing uncontrollable events, including war and terrorism. In this chapter, I will underscore the importance of this perspective by illustrating that in the convergence of 'nature and culture', the power of the Sri Lankan state emerges, eliciting novel forms of governance that are continually justified by the creation of uncertainty and persistent threat.

'Road Map for a Safer Sri Lanka': A preparedness rationale

On 26 December 2004, Sri Lanka was caught off guard. Without warning, the familiar waters of the Indian Ocean receded and returned with a mighty vengeance. The unexpected tragedy elicited an 'if only we were prepared' form of regrettable sadness in Sri Lanka. If only there was a warning system; if only people had known that a retreating sea signals danger. Recent disaster management initiatives have turned these 'ifs' into 'whens'. The disaster management cycle follows a circular rationale wherein disasters are expected potential risks, and thus disaster management becomes risk management.[1] In his analysis of the aftermaths of the September 11 tragedies in New York and of Hurricane Katrina in New Orleans, Andrew Lakoff (2007, 2008) stresses that framing disasters as risks, whether natural or human-made, categorizes them as forms of threat, which in turn, brings them all under a common national security field. Under the purview of preparedness, disaster mitigation solicits new technical band-aids: warning systems, infrastructure management and an overall increase in government management and control. It also enlists the work of various actors such as scientists, urban planners, humanitarian aid and governmental control. The impetus of preparedness efforts is to invoke a continual state of readiness and maximum security of state territory (Lakoff, 2008).

Following the tsunami, the Sri Lankan government formed a Parliamentary Committee on Natural Disasters, whose mandate was to assess the level of preparedness towards disasters. This committee established the Disaster Management Act No. 13 in May 2005 which 'provides for a framework for disaster risk management in Sri Lanka and addresses disaster management (DM) holistically, leading to a policy shift from response-based mechanisms to a proactive approach toward DRM (Disaster Risk Management).'[2]

As outlined in their 'Roadmap for a Safer Sri Lanka', 'safer' primarily entails a reduction of risk, in which assessments of vulnerability, hazard and risk are key in creating a state of preparedness for all types of disasters. This shift away from response and towards preparedness is overseen by the newly created National Disaster Management Centre undertaking the express '... mission to create a culture of safety to reduce the vulnerability of the population to natural hazardous events in the future.'[3]

Initially as an emergency tactic, then President Chandrika Bandarnaike Kumaratunga called for a uniform 200-metre no-build buffer zone along Sri Lanka's coastline. Roundly criticized for making such a blanket border, the buffer zone was then reduced to 100 metres. Now, the buffer zone has been tailored for geographic variances along the coastline. In the east, the buffer zone is 65 metres and even 45 metres in some places. It is dotted with upturned wells, dusty and broken bricks and foundations. Those whose homes were located in the buffer zone were compensated with new homes, even new villages and communities. Relocation and reconstruction have been difficult processes, especially in the east where land scarcity issues prevent movement of families and villages. Even now, more than five years later, some families still await their promised new homes.

The buffer zone and relocation have presented numerous problems. For example, access has been restricted for fishermen and their families whose livelihoods depend on proximity to the sea. Some families now living nearly 14 kilometres from their original homes find themselves in barren and empty government land, with nary a job opportunity or town centre nearby. Some families have been split up in the relocation process, while others have to endure living in poorly and hastily built homes and flats.[4]

In addition, the 2004 tsunami marked a watershed in the use of technology to aid in disaster response and disaster risk management.[5] To this end, one of the main technological efforts the government has undertaken with the assistance of donor countries is to implement a national early warning system. The warning system consists of the construction of multi-hazard warning towers dotting the Sri Lankan coastline – a total of 50 in all. These towers are all connected, that is, they 'communicate with each other' via satellite, from the village, district, and all the way up to the national-level disaster management centres. The main operating centre is located in the Disaster Management Centre in the capital city of Colombo.

The warning towers deliver warning messages to the communities that live around them. When the meteorology department in Colombo gets word of an earthquake or a potential tsunami from their information sources – namely the Japanese Meteorological Association and the Pacific Tsunami Warning Centre in Hawaii, USA – they then contact the operating centre at the Disaster Management Centre. The main centre can either directly send a message to the towers to signal a warning or the district-level coordinators can manually turn the warning sirens on. The messages are recordings, spoken first in Sinhala, then Tamil, and lastly in English. The first message issues a warning. If evacuation is necessary, then an evacuation message will be broadcast. Finally, when the danger passes or if there is no need to worry, a cancellation message will be issued.

Evacuating for the next big tsunami, or how to account for mess in the world

Merely putting up signs and towers, however, is not sufficient for a warning system. As my friend and local disaster management coordinator explained to me, people must properly understand the meanings of the different warnings, which range from tsunamis to cyclones, and emit messages at the different stages stated above. One July, I found myself standing with a crew of disaster management officers under the late afternoon sun in a village on the east coast, staring at the tsunami warning tower. The National Disaster Management Centre wanted to conduct an evacuation drill, in anticipation of the coming solar eclipse that was going to take place in the following days. This solar eclipse was sandwiched between two lunar eclipses, and articles in the newspapers, predictions by astrologers, as well as general gossip and talk pointed towards the possibility of another tsunami or worse, a different and unexpected disaster.

The drumming from the local temple (for it was temple festival season) pounded in the distant background. Earlier we had gone by the busy temple to distribute fliers informing them of the drill and if they heard the warning siren go off, to not be alarmed as it was just a test. In another area of the town, other villagers had been previously selected by the local disaster management coordinator to participate in the drill. Their instructions were to act as if a tsunami was coming and to leave their houses for the 'safe site' as if it were a real emergency.

The siren was scheduled to go off at 3pm and so we found ourselves on the hot road, facing the tower, awaiting, sweating, anticipating the warning message. I had heard it previously on the computer, but never broadcast through the towers. The message was to be dispatched from the main operating centre. Three o'clock came and not a sound. Yet another 15 minutes and still no warning. I sent a text message to my contact at the district-level office, wondering what was happening. 'Just wait', he wrote back. At this point, several of the villagers who also were anticipating a 3pm warning started to slowly trickle out of their homes, turning and looking towards the direction of the warning tower. Some had their emergency bags packed and with them. Others grabbed plastic lawn chairs and sat in front of their homes, wondering and waiting as well. So we waited until at 3:30pm, the local disaster coordinator got a phone call from the national office instructing him to send off a 'local' warning. He promptly grabbed his bullhorn, got on the back of a motorbike and had someone drive him through the village roads with a bullhorn siren. People then began to move towards the 'safe site', a Hindu temple located just outside of town, about one kilometre from the shoreline and a little elevated from sea level. As we journeyed towards the temple, some people were rushing, some were walking, all were carrying bags of some sort. Some were laughing. One woman shouted, 'There is a tsunami!' while she quickly scurried towards the temple. After everyone was gathered at the temple safe site, they were debriefed by the disaster management coordinator. However, most left quickly as they were anxious to get back to the seaside temple to participate in festivities. In casual conversation people relayed to me that they now knew what to do and where to go in the event of another tsunami. They claimed they were happy, because at least now, unlike before, the warning tower was there – despite its malfunctions.

This is not the first time that the warning tower has not properly sounded. In September 2007, when there was a massive earthquake off the shoreline of Indonesia, a frenzied evacuation took place. At the time, one of the early warning towers did not wail its siren. Instead the local disaster management officer climbed up the tower and turned the siren on manually, in spite of the fact that everyone had already evacuated towards the main road, inland. The siren has even gone off at unscheduled times, inciting moments of fear and panic. 'It is there', shrugged Mohammed[6] smiling wryly, when I asked him about the warning tower. 'We [our village] cannot say if it will work one day or the next', he explained. As Najima relayed to me, 'We cannot fully believe in the tower 100 per cent. If it works, that is good, but only God knows if it will work.' Rasheed, the owner of a beachside shop confessed, 'I can sleep better at night because it [the tower] is there. Before, even hearing the sound of the waves, I could not sleep. But now there will be something to warn us.'

The socio-technical system of the faulty warning tower is not so much of an issue. Networks and systems, mediated by 'human' technology always have the potential to go awry, to lack seamless perfection. People need ways to address 'the messy' matters of social life, as suggested by John Law (2004). In his treatise on social science method, he describes the world as full of 'mess' and multiplicity – all of which our research will not be able to account for. This is not a cause for despair. Rather, it is a call to aim our research towards understanding how different realities can coexist and cohere, perhaps relationally juxtaposed to one another. Law illustrates that even with the reality of preparedness and the concern for another tsunami, still other beliefs, critiques and notions articulate with and circulate around what appears to be the overarching institutionalized sentiment of imminent disaster that must be mitigated.

The sentiments of people living near the warning towers signal that they believe something will come. In addition to the staging of several evacuation drills by the Disaster Management Centre, there have been numerous tsunami scares, the most recent ones in the spring of 2010, when Indonesia was again rocked by a devastating earthquake and when destructive tsunami waves hit Samoa. Local astrologers have predicted that the coming years will bring unforeseen catastrophes. False alarms and rumours of tsunamis spread on days with high winds. 'We are always alert', Laxmi and many of my friends and informants told me. She, like many others living near the shoreline, are aware of the slightest shift in wind, the grey colour of the sky, the behaviour of birds. Temple and mosque loudspeakers now double as community warning broadcasters. Alongside and beyond the preparedness rationale of the government, people too consider the possibility of another disaster. The threat of it persists.

The continual threat and constant possibility of a disastrous event guides disaster management, not just institutionally but also in daily life. Surely this is common sense after what was experienced in 2004 when people lost lives, their homes, the deeds for those homes and their life savings. With the acceptance and awareness that disaster may strike again, people weave preparation into their daily lives. It is not due to some unimagined threat produced through fear tactics and propaganda, for in many ways in Sri Lanka, the unimaginable already happened

It is not without some sense of irony that this roadmap to a 'safer Sri Lanka' is being paved on a militant context of over two decades of war and combating 'terrorism'. The efforts to build an 'effective' disaster management system in Sri Lanka in the name of national security parallels some of the very same logics that have come to the fore during the recent years of warring in Sri Lanka. Specifically, the government's approach to so-called terrorism as an issue of national security mirrors the disaster management approach towards 'building a safer Sri Lanka'. That is, government activities and programmes, in the name of national security and safety, hinge upon the ever-present threat of natural disaster or terrorism.

Background on the conflict in Sri Lanka

While the conflict in Sri Lanka has been analysed and written about extensively,[7] I shall give a brief outline of the contours of the conflict, which ended dramatically in May of 2009. Since the early 1970s militant Tamils, which would eventually (not without struggle) become the Liberation Tigers of Tamil Eelam (LTTE), have waged a separatist war against what they felt was a Sinhala-Buddhist nationalist Sri Lankan government. Their grievances were over what they felt was a discriminatory Sri Lankan state that marginalized the rights of Tamils. In this militaristic endeavour they fought for what they saw as their rightful homeland, called *Eelam* (see Figure 10.1), which included the northern and eastern parts of the island.

The beginning of the war is commonly dated as July 1983, when the LTTE killed several Sri Lankan policemen, for which the retaliation was a bloody anti-Tamil pogrom in other parts of the island, perhaps most strikingly in the capital city Colombo. Peace negotiations were attempted several times during the 26 years of conflict. The most hopeful sign of a peace accord happened in 2002, when the LTTE and the government signed a ceasefire agreement that was mediated by the Norwegian government. Despite the agreement, violations continued to be made by both the LTTE and the Sri Lankan government. The LTTE was criticized for its continued conscription of child soldiers and assassination and abduction attempts. They also suffered from internal strife when the Eastern faction of *Eelam* broke away under the auspices of a former LTTE leader, Karuna.[8]

By the time the earthquake that generated the massive Indian Ocean tsunami reached Sri Lanka's shores in 2004, the ceasefire agreement was tenuous at best. There were hopes that the widespread destruction might lead

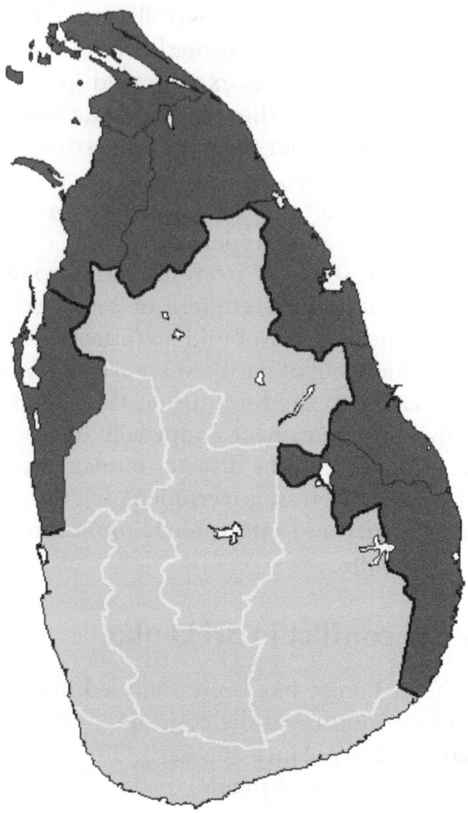

Source: www.wikipedia.org

Figure 10.1 *Map of Sri Lanka and Eelam*

to a unified effort to rebuild the nation. And while the tsunami did provide a temporary respite to the soured and volatile relations between the LTTE and the Sri Lankan government, unfortunately, disputes over the Post-Tsunami Operating Mechanism, an aid-sharing agreement which was never executed, seemed to cast the final blow to the ceasefire. By 2006, the LTTE and the government were engaged in low-scale warfare and by 2008 January, the Sri Lankan government declared the ceasefire officially null.

The government declared 2008 as a year for war in Sri Lanka, and with force and might, the Sri Lankan army slowly began to rout the LTTE from their territorial strongholds in the north and north-eastern parts of the island. By 19 May 2009, the weakened forces of the LTTE were finally defeated. The government's complete takeover of the LTTE's northern territory and the death of the LTTE's ruthless leader, Villupillai Prabhakaran, signalled a final defeat for the insurgent group – at least in Sri Lanka[9] – and prompted the government to boast that Sri Lanka is the only country in the world to have defeated terrorism.[10]

However, it is not as if an easy quiet peace has come to reign over Sri Lanka. The reordering of the nation perhaps comes at the expense of a constant vigilance. As quickly as the green plants grow back over the broken foundations of homes, as sure as the beautiful ocean which once washed red with death and violence breaks on the shore, engagement with the past hovers uneasily over daily life in the east. Another tsunami, another so-called terrorist attack – the uncertain future bears the thick scars of the nation, which do not heal quickly or without difficulty. How do realities and experiences merge and emerge amidst national orderings, out of the past and into the future?

By the end of 2008, the Sri Lankan government's defence budget soared to record-high levels, fulfilling earlier projections of making 2008 the year for war. After the government declared victory over the LTTE, the size of the military was increased by 50 per cent.[11] In October 2009, five months after the end of the war, the Sri Lankan parliament approved a 20 per cent increase of the military budget after the government claimed that security forces need strengthening.[12] Parliament also extended Sri Lanka's State of Emergency, which has been in effect off and on over the last 30 years of the conflict.[13]

This is to say nothing of the quarter of a million Tamil civilians interned in displacement camps[14] in the now defeated LTTE stronghold in northern Sri Lanka, who struggle to get enough food, water and proper shelter. They are being held against their will by the government, living within barbed wire borders of the camps that are guarded strictly by the military. Despite international human rights groups and local social activists clamouring for the release of these Tamils, the Sri Lankan government suspects that Tamil militants still lurk amongst civilians and therefore insists that the interned civilians must go through an extensive screening process. And while it is true that land must still be cleared of mines, people are still not allowed to freely live and stay with their families outside the camps and outside the mine-ridden areas.

Intellectuals and human rights activists are not the only ones decrying the unfolding events of the end of the war. In an eastern coastal town, where some of the worst violence and fighting occurred during the conflict, people commented on my seeming naivety when I probed them about the end of the war: 'Who told you the war was over?' asked Fawzi, a Muslim man, living in a new tsunami housing complex. 'Have the food prices gone down? Are there new jobs? If they [the government and the LTTE] want to start up the war again whenever they want, they can.' Sitammah, an elderly Tamil woman asked me, 'Are the checkpoints gone? When I can travel around this country without having trouble at the checkpoints, then we can say the war is over. When I do not have to worry about my grandsons and nephews not returning home when they go out for work or to visit friends and relations, then maybe we can say that some change has happened.' Still others, like Padmini, a young Tamil woman, expressed in hushed tones that she was not totally happy that the war was over, and that she felt sad that the leader of the LTTE had been killed. Sitammah reiterated that 'inside she was sad', but she said she could not freely express this sadness out of fear that she would be taken into police custody or beaten or, even worse, both. Against televised images of a triumphant nation – shooting cannons and firecrackers in a frenzy of nationalistic pride and unity – the division where I was living at the end of the war was a virtual ghost town, save for an occasional Sri Lankan flag hanging over a shop entrance.

The threat of arrest or physical violence for expression of dissent or unhappiness about the current regime and the end of the war is further reinforced by the ever-present Special Task Force (STF), a special branch of the police that has military-like powers during a state of emergency. Every few weeks, the STF conduct what is called a 'round-up', a form of border control and monitoring process of who goes in and out of homes and villages. 'We are quite used to it', my friend Parvati explained to me. 'As long as we have all our documents and have registered all of our family members with the police, the STF will not give us any trouble when they are doing their household checks.' The STF has become a familiar presence in these areas. It was not uncommon to see them out and about, patrolling, armed with AK-47s, entering homes, and then taking a rest at a local tea stall.

Indeed, as an interview with the commanding officer of the STF revealed to me, 'The war is over, but terror is there.' As I would find out, the division in which I was living and researching was deemed a 'vulnerable' area, and it only made sense to increase the number of troops as a measure to counter terrorism and 'secure the situation'. This was confirmed, for in the weeks right up to my departure in early 2010, the STF had begun to build a new camp.

For the Sri Lankan government, the threat of another LTTE resurgence, or as they prefer to label it 'terrorism', continues unabated. In contrast, for some communities and folks living in 'vulnerable' areas, the threat to their safety and security does not necessarily stem from the LTTE but rather from government and/or state-sponsored violence, punishment and arrest (although the LTTE most certainly have a violent record of kidnapping, child conscription and killing as well). The threat of the government's power is justified by the persistent threat of terrorism and the need for national security.

Disasters as nature-cultures

In Sri Lanka, as I have illustrated, the tsunami articulates with Sri Lanka's civil war; new practices of national preparedness in disaster management efforts coincided and engendered ways in which 'terrorism' would also be managed – all in the name of national security. After the tsunami, hopes of 'rebuilding' the nation were soon dashed by the resurgence of violence and war. Now, in the aftermath of the civil war, doubts linger about so-called 'peace' and life still persists under the prospect of another disaster.

I include the failure of the warning system and side-comments and criticisms of Sri Lankans to point out that despite the overwhelming sense of insecurity created by disaster and terror management by the Sri Lankan state, other realities can also cohere and abound with those that seem to loom large. Disasters move and morph and acquire different meanings in different locales. They are articulated by the intersections of different forces of nature, law, culture and infrastructure. An approach to understanding and studying complexity is to consider a world of multiplicity (Mol and Law, 2006). This approach suggests that we recognize the possibility of coexistences in a *single* moment – a variety of orders, logics, frames, styles, discourses that relate to each other, that mediate and transform each other, that perhaps fit together neatly, or perhaps contradict each other, all without breaking things down into dichotomies and simplifications. John Law encourages taking up the messy matters of social life as a fruitful journey in which to question our practices and methodologies of studying the world (Law, 2004). Understanding the complexities of rebuilding and reordering the Sri Lankan nation is a way of examining how things can come to cohere – not always felicitously, not easily or cleanly, but always relationally.

Notes

1 The geneology of this shift in rationale has been cogently outlined by the works of Stephen Collier, Andrew Lakoff and the Anthropology for the Contemporary Research Collaboratory. See, for example: Lakoff, 2007, 2008; Lakoff and Collier, 2008.
2 Final Report of Sri Lanka Parliament Select Committee on Natural Disasters, May 2005, www.srilankanparliamentonnaturaldisasters.org/Key%20Recommendations.htm, accessed 3 March 2010.
3 Final Report of Sri Lanka Parliament Select Committee on Natural Disasters, May 2005, www.srilankanparliamentonnaturaldisasters.org/Key%20Recommendations.htm, accessed 3 March 2010.
4 For a discussion on the securitization of fear and the production of insecurity through the implementation of the buffer zone in Sri Lanka, see Hyndman, 2007.
5 I detail some of the technological undertakings after the tsunami in Sri Lanka in 'A Safer Sri Lanka? Technology, Security and Preparedness in Post-Tsunami Sri Lanka', in M. de Alwis and E. Hedman (eds) (2009) *Tsunami in a Time of War*, ICES, Colombo.
6 All names in this chapter are pseudonyms to maintain anonymity of the informant.
7 For more information and analyses regarding the conflict in Sri Lanka, see, for example: Abeysekera and Gunasinghe (1987); Spencer, J. (1990); Jeganathan, P. and Ismail, Q. (1995); Winslow, D. and Woost, M. (2004).
8 The breakaway was also assisted, although at first covertly, by the Sri Lankan government. Karuna, a former LTTE colonel, now has the position of Non-Cabinet Minister of National Integration and Reconciliation.
9 The LTTE is now in the process of reorganizing itself politically into a 'transnational government' on a democratic and peaceful platform. For more information: ltteir.org/.
10 The end of Sri Lanka's civil war was mired in controversy. The final numbers of civilian and military deaths published by the UN and organizations such as the International Crisis Group have been disputed by the government. There was also an attempt to begin a human rights inquiry into the government's war tactics, but the move was defeated with Sri Lanka garnering the support of countries such as Iran, China and North Korea.
11 'Sri Lanka's Expanding Peacetime Army', www.defence.lk/new.asp?fname=20090518_12, accessed 5 March 2010; news.bbc.co.uk/2/hi/8121385.stm, accessed 5 March 2010.

12 www.lankabusinessonline.com/fullstory.php?nid=2048117281, accessed 5 March 2010.

13 www.tisrilanka.org/?p=4014#more-4014, accessed from Transparency International 5 March 2010.

14 In January 2010, the interned civilians were given special passes that permitted them a 13-day leave to visit relatives, on the condition that they would return.

References

Abeysekera, C. and Gunasinghe, N. (eds) (1987) *Facets of Ethnicity in Sri Lanka*, Social Scientists' Association, Colombo, Sri Lanka

Bankoff, G. (2001) 'Rendering the world unsafe: Vulnerability as Western discourse', *Disasters*, vol 25, no 1, pp19–35

Blanchot, M. (1986) *The Writing of the Disaster*, translated by A. Smock, University of Nebraska Press, Lincoln, NE

Bowker, G. C. and Star, S. L. (1999) *Sorting Things Out: Classification and Its Consequences*, The MIT Press, Cambridge, MA

Callon, M. (1986) 'Some Elements of a Sociology of Translation: Domestication of the Scallops and the Fishermen of St. Brieuc Bay' (abridged version 1998), in M. Biagioli (ed) *The Science Studies Reader*, Routledge, New York

Dawdy, S. (2006) 'The taphonomy of disaster and the (re)formation of New Orleans', *American Anthropologist*, vol 108, no 4, pp719–730

Fortun, K. (2001) *Advocacy after Bhopal: Environmentalism, Disaster, New Global Orders*, University of Chicago Press, Chicago, IL

Jeganathan, P. and Ismail, Q. (eds) (1995) *Unmaking the Nation: The Politics of Identity and History in Modern Sri Lanka*, Social Scientists' Association, Colombo, Sri Lanka

Hewitt, K. (1997) *Regions of Revolt: A Geographical Introduction to Disasters*, Longman, Edinburgh

Hyndman, J. (2007) 'The securitization of fear in post-tsunami Sri Lanka', *Annals of the Association of American Geographers*, vol 97, no 2, pp361–372

Lakoff, A. (2007) 'Preparing for the next emergency', *Public Culture*, vol 19, no 2, pp247–271

Lakoff, A. (2008) 'The generic biothreat: Or, how we became unprepared', *Cultural Anthropology*, vol 23, no 3, pp399–428

Lakoff, A. and Collier, S. (2008) 'Distributed preparedness: The spatial logic of domestic security in the United States', *Environment and Planning D: Society and Space*, vol 26, no 1, pp7–28

Latour, B. (1993) *We Have Never Been Modern*, translated by C. Porter, Harvard University Press, Cambridge, MA

Law, J. (2004) *After Method: Mess in Social Science Research*, Routledge, Abingdon, Oxon

Masqualier, A. (2006) 'Why Katrina's victims aren't *refugees*: Musings on a "dirty word"', *American Anthropologist*, vol 108, no 4, pp735–743

Mol, A. and Law, J. (2006) 'Complexities: An Introduction', in A. Mol and J. Law (eds) *Complexities: Social Studies of Knowledge Practices*, Duke University Press, Durham, NC

Oliver-Smith, A. (2002) 'Theorizing Disasters: Nature, Power, and Culture', in S. Hoffman and A. Oliver-Smith (eds) *Catastrophe and Culture: The Anthropology of Disaster*, SAR, Santa Fe, NM

Quarantelli, E. L. (1985) 'What Is Disaster? The Need for Clarification in Definition and Conceptualization in Research', in S. Solomon (ed) *Disasters and Mental Health: Selected Contemporary Perspectives*, US Government Printing Office, Washington DC

Smith, N. (2006) 'There's no such thing as natural disaster', *Understanding Katrina*, SSRC, understandingkatrina.ssrc.org/Smith/, accessed 5 March 2007

Spencer, J. (ed) (1990) *Sri Lanka: History and the Roots of Conflict*, Routledge, London and New York

Winslow, D. and Woost, M. (2004) *Economy, Culture, and Civil War in Sri Lanka*, University of Indiana Press, Bloomington, IN

Wisner, B. (1993) 'Disaster vulnerability: Scale, power, and daily life', *GeoJournal*, vol 30, no 2, pp127–140

11
Response and Recovery in the Remediation of Contaminated Land in Eastern Germany

Alena Bleicher and Matthias Gross

Contaminated sites as 'creeping disasters': Beyond risk-oriented thinking

There are as many as 500,000 contaminated sites in Europe, some 20,000 of which can be classified as so-called megasites due to the complexity of soil and groundwater contamination found on them. Megasites pose a range of tricky technical and management challenges with regard to risk and re-use. Situations involving a lack of economic incentives, ongoing health threats, and often unknown chemicals shifting beneath the surface prompt us to refer to such contaminated sites as slow, or creeping, disasters.[1] However, as Uriel Rosenthal (1998, p153) has nicely phrased it, 'creeping disasters can, of course, burst into the open at a specific site, on a specific moment'. The present chapter highlights certain strategies applied to such sudden bursts of contaminants during the process of recovery of a complexly contaminated site located in the former East Germany. These strategies were related to having to cope continuously with ignorance regarding multiple contaminant sources and plumes from previous industrial activities. Any analysis of these processes of dealing with the unknown, we contend, demands clear recognition of and communication about the limits of knowing. This can be seen as an important and yet so far unexplored variable in remediation processes on contaminated land per se.

One of the main challenges posed by such brownfield sites is that in their current form they appear quite green and natural: it seems as if they have already been recovered by nature. Furthermore, much knowledge regarding previous pollution has been lost. Worse still, many contaminants cannot be detected, even with the most sophisticated equipment (Freudenburg, 1997; Picou, 2009). Nevertheless contaminants originating from industrial, mining or military activities on these sites may accumulate to such an extent that they hamper soil functions, pollute groundwater and surface water, and thus threaten drinking water supplies and aquatic ecosystems. Often such 'creeping disasters' only come to light (again) by accident, when contaminants are mobilized by floods, as described by Frickel and Vincent (Chapter 1), or by construction activities on site or close to the site, as we discuss in this chapter. Thus regions subject to these types of creeping disaster have to deal with a range of recovery challenges. These include minimizing adverse impacts on ecosystems and human health and creating incentives to encourage (new) companies to engage with a process of economic restructuring – in other words, making the re-use of brownfields attractive to investors, even if remediation of a site is more expensive than locating on greenfield land.

One of the effects of dealing with contamination and the complexities of remediation over the last three decades has been growing public awareness with regard to reducing land use (Edwards et al, 2005; Thornton et al, 2007). Germany's stated aim to restrict land consumption to 30ha per day by 2020 (Bundesregierung, 2002) has made the re-use of industrial brownfields an important issue, one which increasingly influences decisions regarding land use and location policy. However, in many cases involving the restoration or revitalization of industrially altered landscapes, public debate about the limits of knowledge has been avoided on the assumption that it would undermine the public's confidence in science and policy making. This has meant that ignorance and its material consequences have not been discussed. Only known and observable 'facts' are presented in policy and risk assessments, while the results of risk assessment itself are held to be reliable knowledge, even though they are actually probabilistic estimates. This aspect points to the general limits of the field of risk studies (Hubbard, 2009) and has even led to a fundamental questioning of risk as a meaningful sociological concept over the last 10 or 20 years (Grundmann, 1999; Mythen, 2005; Green, 2009). Among the many concepts of risk that have been formulated, perhaps the most encompassing sociological definition stems from Niklas Luhmann (1993), who defined risk as the attribution of (usually) negative consequences of a decision to a decision maker. This, of course, raises the empirical question of what would *not* constitute a risk-based decision. Judith Green sums up her unease regarding this and other sociological concepts of risk, especially in the field of the sociology of health and illness, as follows:

> *From the perspective of a jobbing sociologist, interested in questions around health and illness, particularly in how people deal with misfortune and make decisions about health in contemporary society, research and theory that is self-consciously 'about risk' appears to offer progressively less that is empirically or theoretically insightful.* (Green, 2009, p494)

We agree with this observation and contend that, in many other cases, when scholars talk about risks and uncertainty, it would be empirically and theoretically useful and more meaningful to frame decisions, utterances and practices according to different shades of ignorance.[2] Our view, therefore, is that decisions made in situ during clean-up processes on contaminated sites can best be framed empirically as decisions made in the face of ignorance.

Decision making and the acknowledgement of ignorance

For most practitioners in environmental management, it is clear that scientific knowledge as well as research in general is always limited by ignorance. This highlights the importance of developing new forms of governance. This task is by no means easy, however, given that, in order to be able to act, the actors involved need to agree on what is not known and take this into account in future planning. They need to decide to act in spite of (sometimes) well-defined ignorance, or what more recently has been called 'nonknowledge', a term used to denote the possibility of becoming knowledgeable about one's own ignorance.[3] While the term 'ignorance', with its occasional meaning of actively ignoring something, is quite ambiguous, 'nonknowledge' refers more precisely to a specific form of the unknown. Ignorance, as a generic term, is defined as knowledge about the limits of knowing in a certain area. More generally, attempts to grasp notions such as ignorance are spurred by the well-known paradox that whenever new knowledge arises, the perceived amount of nonknowledge increases at least proportionally to it, since every state of knowledge opens up ever more intimations of what is not known.

We seek to contribute to the debate by looking more closely at how actors appear to deal successfully with the fact that one never knows what to expect in restoration and remediation processes on contaminated sites. Usually, the remediation of contaminated sites in both active as well as deserted industrial areas is crucial for hazard prevention and for municipal business development. When contaminants are discovered, the actors concerned are forced to deal with substances whose removal entails high costs. However, knowledge about contaminants often reveals more knowledge about what is unknown, since many of the contaminants are very hard to detect in preliminary soil and groundwater investigations; indeed, more often they are only discovered in situ during a clean-up process. In a sense, then, dealing with ignorance becomes 'normal': one is forced to deal with the unknown in such a way that a foundation is established for acting even despite the limitations exhibited in risk assessments and scientific knowledge. Whereas faith in total control and complete knowledge of ecological systems and social processes implies an ability to act only when everything is known in advance, a more experimental approach makes it possible to accommodate a variety of factors in spite of gaps in knowledge.

Using an example from a major research programme related to the remediation of regionally contaminated aquifers in eastern Germany, we shall focus on some of the stages in a clean-up process undertaken in the industrial area of the municipality of Weissandt-Gölzau in the German state of Saxony-Anhalt. We illustrate the strategies for dealing with ignorance used by the heterogeneous actors involved, focusing especially on how they handled surprises and renegotiated previously agreed plans. Nonknowledge, defined as the acknowledgement of ignorance, can be identified in documents and interview sequences at those points where analysis reveals a decision to act after or in light of an event that was communicated as 'surprising' and where the actors became aware of the fact that they did not know enough about something. Theoretically speaking, nonknowledge exists when an actor knows that, in reference to a specific research question, not enough knowledge is available and he or she then defines this knowledge as crucial for decision making. However, the actor is able to name the point of reference of what is not known. During our study, we found that a major part of the actors' strategy was to expect surprising events, which then acted as a trigger for becoming aware of their ignorance.[4]

One hundred years of creeping contamination: Remediation of an industrial site

The municipality of Weissandt-Gölzau is located in *Mitteldeutschland* (central Germany), a region in which, by the end of the 19th century, coal mining had become the main industry. In the 1920s chemical industries based on brown coal refining were established there in order to distillate fuel and tar, along with carbonization plants. As was common practice at this time in all industrialized countries, the residual products from these industrial processes were dumped close to the production facility without any special treatment. In addition to this, all flammable liquids and chemicals were literally dumped into the ground during the period of allied bombing at the end of World War II in order to prevent the factories from being burned down. One long-lasting effect of these activities is that the region still has to deal with the – often forgotten – consequences of these activities. As far as contamination is concerned, things did not get any better after 1945. During the 1950s and 1960s processing activities in Weissandt-Gölzau switched from coal to oil. The production of polyethylene film began, accompanied by a mechanical engineering facility. Since 1990, the commercial enterprises located in Weissandt-Gölzau have been successfully restructured in the course of the wider socio-economic changes occurring in the region after the fall of the Berlin wall. This is quite an unusual phenomenon, not only in this region but also for most areas of the formerly socialist part of Germany, which suffer disproportionately from environmental damage and population decline caused by deindustrialization.

Today, Weissandt-Gölzau hosts two major companies active on the global market, as well as dozens of smaller ones. The two main companies are a polyethylene film production company and a mechanical engineering factory. In the restructuring and consolidation phase, which lasted about ten years, not all parts of the industrial area were used: some were left deserted as brownfield sites. It was only during the early 2000s that the larger companies on the site began to expand onto the contaminated brownfields in the vicinity. Due to existing knowledge regarding the site's former industrial use (and almost certainly also mistrust regarding previous environmental practices), private companies located there were keen to establish the status of contamination. Since much of the contamination originated from the early chemical industries of the 1920s, records of accidents and waste disposal leading to contamination were very rare. In the course of economic restructuring after World War II, during the 1950s and 1960s and especially after 1989, many parts of the industrial facilities were rebuilt, buildings were demolished, sinks filled and other buildings constructed in the same place. Very little documentation (archive materials or maps) about these changes exists today; most of the information that would otherwise have been available disappeared when new administrative structures were established after 1990. Many of the events can only be reconstructed from interviews with the people who lived in the area long enough to remember some of the activities on these sites and who have a kind of 'mental map' of the area, as described by Henderson in Chapter 5. Between 1990 and 1996, local enterprises contracted specialist engineering companies to investigate water and soil contamination in some parts of the area. No significant contamination was found on these tracts of land. However, experts in local and regional agencies were aware that further investigation of the remaining tracts would be needed. Due to a lack of investment in public infrastructure in the industrial area, the political representatives of the municipality decided in 2002 to improve the existing roads and walkways as well as the sewage and rainwater drainage systems in order to make the site more attractive for investors. Due to existing knowledge about the area's industrial history and a number of unexpected encounters with contaminants during site developments since 1990, both the municipality and the relevant regional and district authorities were aware that construction activities in the area would have to deal with contamination of unknown dimensions.

One of the most important activities at the site was a project carried out by the municipal authorities aimed at improving the infrastructure. It was the largest investment project undertaken to date and was accomplished by a heterogeneous group of actors. Although investigations into contamination were carried out in advance, diverse types of hitherto unknown contaminants were discovered not long after construction work began (Gemeinde Weissandt-Gölzau, 2007). Given the site's history and the actors' anticipation of surprises in relation to finding contaminants, one could say that the actors involved were aware of their own ignorance (contamination of unknown dimensions): they expected the unexpected. As the principal manager of the contracted local construction company explained: 'Sometimes different activities took place simultaneously. Sometimes, though, we didn't know what to plan for, because there were no studies we could build on.' This awareness, of course, influenced each decision made by the stakeholders involved.

Reframing nonknowledge through stakeholder partnership

The institutional context for remediation projects and stakeholder decisions in Saxony-Anhalt is determined by national as well as federal state regulations.

The most important law in Germany concerning the remediation of contaminated sites is the Federal Soil Protection Act of 1998. This law clearly defines the overall principle governing the remediation of contaminated areas, namely the 'polluter pays principle'. It sets specified limits for hazardous substances, describes the procedure to be followed in remediation projects, and names the institutions to be involved. Federal state and regional authorities are responsible for overseeing risk prevention and remediation. In the eastern German federal states, an additional regulatory body exists at the national level, the *Landesanstalt für Altlastenfreistellung* (LAF, State Authority for Exemption from Residual Pollution Liability). The purpose of this authority is to promote the remediation of environmental pollution resulting from former industrial activities and, in so doing, to encourage economic investment. The LAF is the only body of its kind in Germany and other European countries and decides on financial support in coordination with the relevant environmental administration. In addition, a fund was created to provide financial support for such projects. In Saxony-Anhalt the fund is administered by the LAF.

In general, the removal of specific contaminants gives rise to enormous costs which cannot be precisely foreseen due to often limited knowledge regarding the nature of the contamination. In the case of the Weissandt-Gölzau project, the unknown dimensions of contamination, combined with a limited budget and a strict project time frame determined by public funds, necessitated the involvement of heterogeneous stakeholders with varying financial and professional resources. A wide range of stakeholders, institutions and their representatives brought in not only crucial financial resources but also a variety of interests, practices, working styles and professional cultures. In the following, we describe the six most important stakeholder groups, focusing on the role played by each in cleaning up the site, along with their internal interests, financial resources and professional expertise. Our account is based on the actors' descriptions as provided in the interviews. Our aim is to illustrate some of the organizational strategies involved in coping with unexpected findings regarding contamination.

First, there is the municipality, a local authority with few financial resources and little specialized expertise. The local authority is responsible for the maintenance of the infrastructure, which was the major driving force in undertaking the project and promoting site development. The overall interests of the local politicians have to do with maintaining the (administrative) functionality of the municipality and the various practical aspects of public life. Both issues are heavily dependent on the number of inhabitants in the area and the local economy: the income of municipalities in Germany derives from income taxes and from allocations of public money based on the number of inhabitants. For this reason, municipalities are extremely keen not only to keep existing commercial enterprises within their town limits but also to attract more to the area in order to foster employment and population growth. Despite the relatively low unemployment rate (5 per cent in Weissandt-Gölzau itself, compared with around 15 per cent in the state of Saxony-Anhalt), municipal revenues have been too low to finance new infrastructure and remediation measures without external funding. In addition to this, local administrations rarely have specialists at their disposal who are able to deal with complex issues like the remediation of contaminated areas.

The second stakeholder in the development project was the State Authority for Exemption from Residual Pollution Liability (LAF, see above), which provided funding as well as professional expertise. LAF's mission is to support remediation measures in order to minimize environmental health hazards and to support economic development. Given that contamination problems were assumed to exist and that municipal funds were limited, it was obvious that this authority would necessarily be a partner in the project. However, LAF has neither the funds available for building infrastructure nor the necessary professional expertise to oversee construction projects.

The third major player in the project was the *Bitterfelder Qualifizierungs- und Projektierungsgesellschaft* (BQP, the Bitterfeld Qualification and Project Management Association), a management firm called in to oversee the infrastructure and remediation project. Based in the project region, BQP was contracted by the municipal authority to organize and manage the entire infrastructure project, from planning through to remediation and construction. Its main interest lay in implementing the project quickly and efficiently, which included the effective coordination of all the partners (construction companies, financing partners, regional administration etc.), achieving the project goals, and finishing the project on time. Beyond this, the company's aim was to generate profits and to earn positive references for further assignments in the region. The company has a long-standing history of managing construction projects in the region.

The fourth stakeholder group in the project, of course, was the group of owners of the respective tracts on the site, who were hoping to sell off their pieces of land at a profit. In most cases, the representing organization was a federal state company, the *Mitteldeutsche Sanierungs- und Entsorgungsgesellschaft* (MDSE, the Central German Association for Remediation and Disposal). MDSE owns many particularly heavily contaminated areas in the region, many of which had belonged to formerly state-owned enterprises. Generally speaking, it is more beneficial to sell land to investors than to local municipal authorities, since the latter usually have only limited financial resources. In the case of Weissandt-Gölzau, the MDSE agreed to sell their property to the municipality (for the construction of streets and a sewage system) because it, too, was keen for the project to be completed successfully.

The fifth group of stakeholders is that of the administrative authorities. The role of local and regional authorities was to grant permission for various on-site activities and to monitor progress. The state environmental agency sought to achieve an improved environmental situation, the building authority wished to have a better infrastructure system put in place, and the business development agency was hoping that more jobs would be created. Some of these authorities had the legal power and professional expertise to support the project manager in obtaining public funding.

The last group of stakeholders comprised the local companies that were not directly involved in the project. At the beginning of the project in 2005, most local companies pursued their own individual interests – production, expansion, profits – while only a few seemed interested in the development of the wider community. Even the idea of renewing the infrastructure in the industrial area was perceived with some suspicion. However, as soon as the local companies had been exempted from all financial liability, they began to give the project nominal support, albeit largely by paying lip service to it.

Many more actors than those named above were involved in this infrastructure project. These were generally contracted to do particular kinds of work: *engineering companies* planned the infrastructure and evaluated soil quality, a *consultant* carried out investigations and evaluations to elicit detailed knowledge of the contaminants present, and a *construction company* removed contaminated material. Although the overall decision making occurred in a top–down fashion, from the municipal level to project manager BQP and LAF, and from there to different contractors and subcontractors, it was the subcontracted workers and engineers on the ground who were allowed to make the initial decision to continue or to stop working when contaminants were found unexpectedly during the clean-up process.

The situation that prevailed in the region gave rise to a shared notion that the key to keeping people in the region was to improve the environmental situation and to foster economic development. Both time pressure and the knowledge that a further decline in population would worsen the overall situation in the region helped those involved to put their conflicting interests on hold. All the parties to the project knew that they might have to raise additional funds themselves, but they all also knew that doing nothing could easily mean less economic development, higher rates of unemployment, and further population decline. In almost every interview we conducted and every weekly planning meeting we attended, this was mentioned as being a crucial issue. Thus there was a recognition that dealing successfully with nonknowledge is an unavoidable reality in the process of creating a sustainable future for a region. The cooperation among the stakeholders that existed on the basis of this insight was fundamental to the success of the project.

Nonknowledge as opportunity

After funds had been approved by LAF in the autumn of 2005, the first stages in the remediation project were embarked upon in January 2006. The project was divided into several subprojects. Each of these operated according to the following scheme: (1) systematic investigation by the consulting company; (2) planning by the engineering company in cooperation with research institutions; (3) building application and building permission from the administration; and (4) construction. Although these stages were designed to build on one another, they were generally undertaken almost simultaneously on account of the prescribed time frame and the pressures of implementation. The task of each company was defined precisely. All the companies carried out their work in the area at the same time, and all the actors involved knew that they would have to cooperate if there were any surprising encounters with contaminants

– surprises anticipated by all the actors involved in one way or another. Given the limited time frame and budget of the project, the actors involved knew that they would have to make decisions based on nonknowledge: no risk assessments were available because the key parameters were unknown. Waiting for 'proper' knowledge would have entailed exceeding the project time limit and frittering away the opportunity to attract new investments. The following statement from one state authority official – of whom one would perhaps least expect that he would speak openly about the limits of his institution's knowledge – indicates a clear awareness that something has to be done, and that waiting for accepted knowledge and risk assessments would be the worst possible decision:

> *In such a complex situation we cannot say exactly what type of measures we need in order to take action against these hazards. We just don't know, although [name of an engineering company] has been conducting research on the issue since 2003. Basic research done by [name of academic research institute], with its more technical aspects, has also been integrated to ensure that the results help us to evaluate different variants. But we still can't say exactly what's going on with the groundwater. And yet we need to do something – don't we? What is our yardstick for doing something? This [i.e. not knowing] doesn't mean that we can just stop the process and tell everyone to wait until we have a risk assessment.* (Interview February 2007)

A further consequence of the need to act was that the actors involved had to expect not only that surprising things might happen but that things could go wrong. Thus the path of experimentation appeared to be a feasible way to move forward. In this spirit, the reconstruction of the first road began in March 2006 and proceeded as planned. The only problem cropped up in November 2006, when several old electric cables were found. However, their removal posed no great difficulty. Renovation of the second major access road began in July 2007. Difficulties arose here some two months later when the bulldozers' progress was blocked by liquid tar. Due to the toxic fumes given off by the evaporating tar and the consequent risk to human health from coming into contact with this substance, workers and visitors alike had to wear protective clothing during the three months that followed. In both cases, none of the actors had expected to find cables or tar underneath existing roads. When construction of the first of four new roads began, the workers were immediately confronted with metal containers filled with tar, which were hard to date (the largest of these containers, measuring 12m in length and weighing 12t, was found in April 2007). A representative of one of the engineering firms said he regarded such occurrences as a normal part of the remediation of contaminated sites:

> *By now all the actors are used to the fact that on every abandoned industrial site some kind of lack of clarity remains. Not everything is documented. Let me put it this way: this tank [filled with twelve tons of tar and solvents], maybe it didn't even want to be documented. After all, it's amazing where those containers were discovered. There are no documents about them.* (Interview July 2007)

It is interesting to note here that the engineer puts the blame on the tank filled with contaminants *itself*, saying that the tank itself *did not want* to be detected. The suggestion is that it was 'hiding' from the scientists' and engineers' high tech searching devices. In other words, rather than putting the blame on one of the scientific technicians who took the soil samples and did the evaluation, the 'mistake' was attributed – albeit with quite an ironic undertone – to the contaminated material *itself*.

Subsequent to these early discoveries, soil contaminated with tar and solvents was frequently found in crumbling basement foundations originating from the early 20th century. Although the removal of these finds slowed down some of the other activities each time, it was at least possible to find quick disposal solutions for the tanks and metal containers while construction work in other places continued. Since it was well known that such tanks can easily catch fire and explode and that expert knowledge is needed to deal with these issues, specialist companies capable of opening and transporting such tanks had to be contracted at short notice. All these occurrences increased the costs of the project, but the overall process continued despite this. As the principal manager of the local planning company put it:

> *Decisions needed to be made on an ad hoc basis, usually by the construction manager. One cannot plan everything [...] In fact even if the map drawings of the roads and their contaminants are correct, what you find in the border areas is unknown and, in principle, can change every day.* (Interview January 2007)

This statement shows that the actors were aware of ignorance and surprises, and it gives an indication of the main problem with which the actors had to deal: being prepared for sudden changes.

In October 2006 a subproject for the construction of a rain storage reservoir was begun. During the first stages of this project, in November 2006, some common domestic waste was found and the first soil samples removed revealed the presence of small amounts of liquid tar. A few weeks later, when the construction company removed the topsoil, workers reported that to their eyes the subsoil seemed to be 'not normal'. The actors involved knew what they did not know and subsequently specified their ignorance (nonknowledge). They were then able to use this nonknowledge as a basis for further planning and action. An engineering company was consulted, which took samples and analysed the soil, concluding that the entire area was heavily contaminated with liquid tar. This led the actors to ask why there was tar and not just soil. The specialist companies were able to reconstruct where the tar had come from, namely a nearby landfill where residues from the carbonization plant had been stored since the 1920s. In this case, the nonknowledge developed from general ignorance was able to be transformed into knowledge. As a consequence, the construction work continued, but a full layer of top soil over one metre in depth was removed and disposed of at a waste site for hazardous materials. Thus the original plan was changed and later adapted successively to each new situation as it arose; this often went hand in hand with obtaining a new building permit. Commenting on this incident, the representative from the engineering company in charge of the project stated: 'The discovery of tar meant that the whole philosophy of the project had to be rethought. A completely new plan had to be drafted ad hoc.' In a similar vein the responsible person from the municipal administration explained:

> *Construction and cleaning-up has to be understood as a continuous process. They [the workers on the ground] started to remove the material, and when they did so they found that the subsoil was totally different from what we expected. The problem is that tar, which turns to liquid at a certain temperature, makes it necessary to use another type of water barrier for the rain storage reservoir. [...] At times like these we have to act first and get the permits changed later.* (Interview March 2008)

Thus the actors involved often had to act in an ad hoc manner. In so doing, they effectively entered a legal grey area. This shows that the impossibility of knowing everything was taken seriously; in other words, no blame was placed on supporting actors if they had to do something which involved moving beyond the boundaries of what is legal. This is an important point, and it recalls the case mentioned above in which a large tank was not detected by the investigating engineers, the blame for which was put not on a person but on the tank itself. Doing otherwise in either case would have slowed down the overall process significantly and may even have brought the project to a halt.

In general, all the actors involved in the project reacted quickly and flexibly, so that a decision about new methods was able to be finalized in April 2007. Additionally, special geo-membranes and barrages were built into the ground to avoid tar seeping into the reservoir and thus into the surface water system. The actors agreed on what was not known and took this into account in their subsequent activities – that is, they decided to act in spite of nonknowledge. This nonknowledge was then fed into the next stage of the process.

Upon completion of this stage of the construction process in November 2007, much – though by no means all – of the contaminated soil had been removed. Nevertheless, the actors were aware that they could be certain that all contaminants had been removed *only* in precisely those places where they had actually worked with their machines and appliances. As was the case in the 1990s, local authorities today are also aware that contaminated tracts of land remain in the industrial area. However, this knowledge of the unknown is also based on preliminary estimates of how liquid tar might move. Thus there is always the potential that a cleaned up area will be recontaminated. In the next section we take a closer look at the strategies used by different actors to deal with the unknown and at how the unknown influenced their decisions.

Nonknowledge and strategies of coverage

As was the case with the decontamination project in Weissandt-Gölzau, revitalization projects are often undertaken on a tight monetary budget and with tight time constraints, so they cannot be postponed until every last detail of future planning is precisely known. Whenever political and economic actors seek to wait until they are in possession of the ultimate scientific truth, they are effectively misusing science and research as an excuse for political inaction. Whenever scientists and engineers claim to be able to deliver hard facts and genuinely reliable knowledge, they are overestimating their own methodological capacity. Experts' willingness to disclose the limits of knowledge in their communication with other stakeholders can thus be understood as an important part of the process of implementing successful development projects.

As the interviews with actors and the analysis of minutes from planning meetings reveal, the strategy that was adopted to deal with unexpected events was based on precise definitions of what was unknown, thus making it possible to devise flexible, cushioning measures around these unknown factors. The actors' acknowledgement of ever increasing ignorance was not used as a means to prevent the occurrence of a surprise find of contaminants; instead, they accepted that this would occur in any case. What *was* created were the prerequisites for making flexible adjustments, such as in the change of permits (administration), in the change of plans (engineering companies) and, last but not least, a certain flexibility in redeploying capital for remediation. One major factor which helped to generate such an 'experimental attitude' was an explicit agreement on shared objectives made at an early stage of the project. As a key representative of LAF put it: 'Early on you have to get the right people to the table: people who trust each other and who also have some credibility for actors outside the project.' Who 'the right people' are, is defined not only by the (financial) resources of potential partners and their power to influence the project but perhaps as importantly by their expertise and credibility. Expertise is exemplified, for instance, by the long years of experience a company has in dealing with the revitalization of contaminated landscapes in the region. Credibility evolves out of actors' experience of regular and repeated interaction with others in working together to deal successfully with unknown factors. Credibility and reliability seemed to be important preconditions among the six actor groups described above, with their multiple and sometimes contrary interests, so that different professional cultures did not have an adverse impact on the climate of cooperation.

In addition to personal acknowledgements of what was not known, ignorance was also expressed in the contracts made between the actors. Special clauses in these contracts and permits – so-called collateral clauses – make particular mention of the unknown. Collateral clauses are agreements between joint contractors to pool their guarantees in handling a large project. The objectives are that others will also have recourse to the contractor's warranty to continue the project in the face of unexpected changes (German: *Auflagenvorbehalt* and *Nachtragsangebot*). These strategies, which we call 'strategies of coverage', were developed specifically for this site's remediation process.

A further important issue with regard to successfully coordinating such projects is the institutionalization of contacts and information exchange. Although this might seem obvious at first sight, regular consultation of all institutions and actors involved for the purpose of exchanging information, discussing new developments and agreeing on strategies for adapting to new situations is not the norm in many projects dealing with contaminants. The official rhetoric still suggests that an engineering company simply delivers completely reliable results based on science, and that if it does not do so, it is no good. Consultation also implies that all the actors involved must communicate their own ignorance, understood not in terms of a failure but rather as a normal way of dealing with contaminants.

Conclusion

Situations in which contaminants originating from industrial activities that took place up to 100 years ago continue to lie below the ground can be called creeping disasters. We use this term to point to the reality of the long-term conditions and long-term development of such disasters. The approach taken to cope with this phenomenon, which we have described in this chapter, is one that accommodates project activities despite the presence of unknown factors. Seen in this way, dealing with ignorance in the course of the remediation of contaminated sites appears as a normal occurrence rather than as an anomaly or an indication of failure. Put another way, dealing with ignorance is not simply a process of trial-and-error and of learning from failure, as failures suggest that mistakes have been made. Instead, the stakeholders and actors we observed took seriously the impossibility of avoiding ignorance, so that there was no target for blame or 'finger pointing'. Unlike in other recovery projects, many activities in Weissandt-Gölzau seem to include ignorance and nonknowledge explicitly as part of planning. This seems to encourage the development of innovative strategies and to make full use of the potential and resources of the actors involved to achieve a common goal. The actors achieved a state of preparedness, found organizational solutions for cushioning the impacts of surprising events, and thus developed strategies for coping with the unknown built on regular consultations and communication.

Because the actors involved in remediation processes usually have to deal with multiple unknown factors, decision makers are generally hesitant to agree to projects aimed at cleaning up contaminated areas (Kampe, 2000). By contrast, the necessity for action in Weissandt-Gölzau was driven by decision makers who were aware of the generally weak state of the economy in Saxony-Anhalt. This was the main factor behind their willingness to take action: doing nothing was clearly regarded as the worst possible decision. Once other actors became involved, they helped to shape experimental strategies which we, as social scientists, were able to analyse.

Given the enormous number of unknown factors regarding the dynamics of contaminated landscapes, the development of new industries, acting in spite of ignorance by acknowledging nonknowledge will perhaps become even more important in years to come. The integrity of the process in Weissandt-Gölzau was successful for the most part because nonknowledge was used skilfully in the planning and implementation stages and because all the stakeholders were fully integrated. The main prerequisite for success in such processes is that the actors involved should be ready to act and make decisions *in spite of* ignorance. In order to do so in a way that minimizes further uncertainty in decision making, the relevant social networks need to be characterized by what Weick and Sutcliffe (2007) have called 'collective mindfulness'. This is not self-evident. For some it might even appear counter-intuitive. Nonetheless, stakeholders such as the ones we studied in Weissandt-Gölzau know that there are many things that can go wrong, and yet they operate without labelling these occurrences as failures. Perhaps the case discussed here can even be seen as an indication that the paradigm change posited by Collier and Lakoff (2008) – from a rationality of insurance and risk to what they call rationality of 'distributed preparedness' – is indeed taking place. In the case discussed here, the actors (self-) organized to put fast and flexible preparatory measures in place in order to guarantee the continuity of the overall recovery and remediation process. Clean-up processes instigated to deal with such cases of creeping contaminants demand fast and flexible governance structures rather than blame and finger pointing.

Acknowledgements

The analysis presented in this chapter is part of a larger research project on the Remediation of Regionally Contaminated Aquifers (SAFIRA II). As staff members involved in this major project, which is funded by the German Federal Ministry of Education and Research (BMBF), we were in a position to enter the field and make contact with the stakeholders involved. We are grateful for the openness shown by our interview partners in speaking about what they did not know.

Notes

1 For discussions and definitions of 'creeping disasters', see Alexander (2001), Murphy (2009) and Rosenthal (1998).

2 Michael Smithson has been making this point for some 25 years (cf. Smithson, 2010). His voice has largely gone unheard.

3 Nonknowledge is a literal translation of the German *Nichtwissen*. Unlike the term ignorance, nonknowledge points to a symmetry between knowledge (*Wissen*) and knowledge about what is not known (*Nichtwissen*). For a detailed description and

analysis of different notions of the unknown, see Gross (2010). Alternative concepts used to frame the unknown can be found in Hess (2009), Kerwin (1993), Smithson (2010), Thompson (1997) and Frickel and Vincent (Chapter 1).

4 Data for this chapter stem from various sources, ranging from participation in meetings and the observation of activities at the site, minutes of previous meetings, evaluations of archive materials and contracts, and semi-structured interviews with 17 stakeholders. The interview transcriptions were codified using MAXqda, a programme enabling qualitative text analysis (Kuckartz, 2004). We identified a set of recurring themes, which were compared and discussed until a coding scheme was agreed upon. In addition, we used the standard methodology of content analysis to identify text passages belonging to predefined themes and categories, based on the concepts of nonknowledge and ignorance. During the analysis, the coded segments were linked to each other in order to illustrate possible causal or temporal relations between the categories.

References

Alexander, D. E. (1993) *Natural Disasters*, Routledge, New York

Bundesregierung (ed) (2002) *Die Nationale Nachhaltigkeitsstrategie*, Report of the Press and Information Office of the Federal Government of Germany, Berlin, www.bundesregierung.de/, accessed 14 June 2010

Collier, S. J. and Lakoff, A. (2008) 'Distributed preparedness: The spatial logic of domestic security in the United States', *Environment and Planning D: Society and Space*, vol 26, no 1, pp7–28

Edwards, D., Pahlen, G., Bertram, C. and Nathanail, P. (2005) *Best Practice Guidance for Sustainable Brownfield Regeneration*, Report on the RESCUE Research Project, www.rescue-europe.com/download/reports/RESCUE%20Manual.pdf, accessed 27 April 2010

Freudenburg, W. R. (1997) 'Contamination, corrosion and the social order: An overview', *Current Sociology*, vol 45, no 3, pp19–39

Gemeinde Weissandt-Gölzau (ed) (2007) *Weissandt-Gölzau: Seit 130 Jahren Industriestandort*, Official Information Book for Potential Investors, Weissandt-Gölzau

Green, J. (2009) 'Is it time for the sociology of health to abandon "risk"?', *Health, Risk & Society*, vol 11, no 6, pp493–508

Gross, M. (2010) *Ignorance and Surprise: Science, Society, and Ecological Design*, MIT Press, Cambridge, MA

Grundmann, R. (1999) 'Wo steht die Risikosoziologie?', *Zeitschrift für Soziologie*, vol 28, no 1, pp44–59

Hess, D. J. (2009) 'The potentials and limitations of civil society research: Getting undone science done', *Sociological Inquiry*, vol 79, no 3, pp306–327

Hubbard, D. (2009) *The Failure of Risk Management*, Wiley & Sons, New York

Kampe, D. (2000) 'Modellvorhaben Sanierungs- und Entwicklungsgebiete: Versuch einer Zwischenbilanz', *Informationen zur Raumentwicklung*, nos 5/6, pp341–347

Kerwin, A. (1993) 'None too solid: Medical ignorance', *Knowledge: Creation, Diffusion, Utilization*, vol 15, no 2, pp166–185

Kuckartz, U. (2004) *An Introduction to the Computer Analysis of Qualitative Data*, Sage, London

Luhmann, N. (1993) *Risk: A Sociological Theory*, de Gruyter, New York

Murphy, R. (2009) *Leadership in Disaster: Learning for a Future with Global Climate Change*, McGill-Queen's University Press, Montreal

Mythen, G. (2005) 'Employment, individualization and insecurity: Rethinking the risk society perspective', *The Sociological Review*, vol 53, no 1, pp129–149

Picou, J. S. (2009) 'Katrina as a natech disaster: Toxic contamination and long-term risks for residents of New Orleans', *Applied Social Science*, vol 3, no 2, pp39–55

Rosenthal, U. (1998) 'Future Disasters, Future Definitions', in E. L. Quarantelli (ed) *What is a Disaster? A Dozen Perspectives on the Question*, Routledge, London

Smithson, M. (2010) 'Understanding Uncertainty', in R. Bammer (ed) *Dealing with Uncertainties in Policing Serious Crime*, Australian National University Press, Canberra

Thompson, H. (1997) 'Ignorance and ideological hegemony: A critique of neoclassical economics', *Journal of Interdisciplinary Economics*, vol 8, no 4, pp291–305

Thornton, G., Franz, M., Edwards, D., Pahlen, G. and Nathanail, P. (2007) 'The challenge of sustainability: Incentives for brownfield regeneration in Europe', *Environmental Science & Policy*, vol 10, no 2, pp116–134

Weick, K. E. and Sutcliffe, K. M. (2007) *Managing the Unexpected: Resilient Performance in an Age of Uncertainty*, John Wiley & Sons, New York

Afterword

Rachel A. Dowty and Barbara L. Allen

Despite many differences seen when comparing disasters, all disasters share certain characteristics. The levels of risk for certain dangers are re-examined. Response effectiveness is scrutinized and weaknesses are found. Recovery is measured in the long-term through the memories of those affected, and in the short term by the length of time it takes for daily rituals to recommence as closely as they were before disaster struck. Whether the disaster is deemed 'natural' or 'man-made', all disasters begin and end in communities and the social groups, networks and politics that sustain them. As such, lessons learned about risk, response and recovery may vary from community to community and from disaster to disaster, but trends in the dynamics of social groups, networks and politics can be used to form prescriptive generalizations across continents and hazard types.

Lessons on risk

The chapters in this volume highlight many lessons learned by specific communities about risks to which they are particularly vulnerable. Two particularly lend themselves to generalization: (1) that an understanding of locations and topics about which knowledge is lacking can be just as important within a community as understanding what is known, and (2) the built environment holds as many intrinsic risks of hazards as geographic location holds for risks of natural hazards.

First, as Frickel and Vincent (this volume) conclude, knowledge gaps about environmental hazards in communities not only depend upon where samples are collected but also when they were collected and the social conditions that predicated the sample collection in the first place. One inherent part of these social conditions is the institutionalized rationality around which risk assessment takes place. The US Environmental Protection Agency (EPA), for example, focused on post-Katrina risk assessment based on evidence requiring regulatory action, while ignoring contamination when regulatory limit thresholds were not exceeded. Another part of these social conditions highlights the fact that groups who take up actions to ameliorate or improve upon the work of regulatory agencies (like the EPA) also uphold the same institutional rationality that they criticize by focusing on the same questions addressed by the regulatory agencies instead of asking new and unaddressed questions about communities and the environmental risks, or creeping disasters (to use Bleicher and Gross's term – this volume) that they face. Thus, an understanding of why certain knowledge is absent can help risk-assessment teams provide local communities with more substantive testing results and explanations of toxic contamination.

Second, focus on uncertainty with regards to natural hazards tends to play up measurements and statistics such as wind speed and rainfall amounts, which can lead to the same reasoning about the risks of the built environment (for example, buildings can withstand certain wind speeds, are elevated by a certain amount). The folly of such rationale in terms of assessing vulnerability in practice is twofold: it fails to account for social and political circumstances of the people in and around those buildings (such as that covered by Bowden, this volume) and does not open disaster management to ongoing and multifaceted inquiry (as suggested by Woodhouse, this volume). In Gall's words (this volume), '[t]oday's vulnerable populations might not necessarily be tomorrow's vulnerable populations'. Or to use Choi's words (this volume), '[d]isasters move and morph and acquire different meanings in different locales'. Thus, the geographic location of a place is but one small part of an accurate picture of risk. Communities grow into their built environments, and as they grow, social vulnerabilities grow with them. These dynamic vulnerabilities cannot be assessed by taking static measures, and must be taken into account simultaneously with static measures if an effective risk management strategy is to reign in any given area. As summarized by Weichselgartner and Brévière (this volume), 'Katrina and the Elbe river floods revealed how traditional risk-management policy does not sufficiently consider natural hazards as a product of the interrelations between the physical environment, society and the man-made environment'.

Lessons on response

A disaster situation is not the time to start planning, yet because the very characteristics of disaster include the unexpected, planning in the middle of a disaster is exactly what emergency managers must do to some degree. And planning under such conditions challenges the most sound of decision-makers. Lessons covered in this volume regarding response therefore focus on decision-making: (1) culture plays a pivotal role in shaping decisions, and should thus be included in both planning and response stages of emergency management, and (2) institutional transparency and accountability are critical to focus response decisions and maintain solidarity in the face of calamity.

Dowty et al (this volume) describe how a pre-disaster determination of an organization's culture can help identify potential response coordination and communication breakdowns. In the throes of response, such an understanding of one's organization can help employees coordinate different rationales for effective response. Cross-cultural comparisons of disasters can also aid collective learning from disasters, as suggested by Weichselgartner and Brévière (this volume). They highlight the importance of including local citizens in decision-making processes in all stages of disaster, as well as the notion that '[s]eparated responsibilities and unclear competences cause difficulties in coordinating disaster response activities and, therefore, the decentralized management systems need to improve structures of responsibility and knowledge about mechanisms, regulations and possibilities'. Culture can play a pivotal role in demystifying decentralized disaster management, on both organizational and nation-state levels.

Another lesson is that blame for disaster outcomes may only become constructive when coupled with institutional transparency and accountability. Fast and flexible governance structures require these characteristics, as covered by Bleicher and Gross (this volume): 'The integrity of the process in Weissandt-Gölzau was successful for the most part because non-knowledge was used skilfully in the planning and implementation stages and because all the stakeholders were fully integrated. The main prerequisite for success in such processes is that the actors involved should be ready to act and make decisions *in spite of* ignorance.' In a climate of emotionally charged finger pointing, acting in spite of ignorance can be a dangerous strategy, but one that is necessary during disaster responses. When communities and organizations are integrated and transparent to the point where each understands where the authority and, sometimes even more importantly, the lack of authority resides, accountability can be constructively used to make lagging responses more efficient.

Lessons on recovery

Two particular lessons about recovery covered by chapters in this volume include: (1) the role of non-linear evaluative practices, such as cognitive mapping, in both understanding the psychological terrain of displacement and accessing situated knowledge as a means of citizen involvement in disaster planning and post-disaster mitigation, and (2) the importance of coordinated participation, intervention, and oversight from both community residents and authorities to avoid the replication and magnification of social and political inequities in neighbourhood planning and repatriation.

Kathryn Henderson's research (this volume) working with post-Katrina refugees demonstrates the phenomenological nature of space and place as it is held in the memory of displaced residents. Through a series of exercises whereby people actually mapped their former communities, she was able to access the kind of 'spatial and navigational knowledge that is tacit, situational and indexical' towards documenting multiple and conflicting meanings of homeplace in traumatized flood victims. The policy implications of this have many layers. Resident mapping can be used as a user-friendly tool for participatory repatriation and rebuilding. It is one way to access some of the complexities of the lived environment, what Choi (this volume) calls 'a variety of orders, logics, frames, styles, discourses that relate to each other, that mediate and transform each other, that perhaps fit together neatly, or perhaps contradict each other, all without breaking things down into dichotomies and simplifications'. The use of a more resident-generated home-view, such as cognitive mapping, may enrich the more quantitative, linear kinds of fact collection, such as surveys, that normally constitute post-disaster data collection.

Finally, it is important to recognize the significance of social and cultural norms in planning and facilitating repatriation and rebuilding after disaster. While outside expertise and resources are desperately needed by those who have lost their homes and livelihoods, just as important are inclusive and participatory processes by which those resources are allocated. According to Barrios (this volume), '[t]he rigid application of expert knowledge has, time and time again, proven to threaten environmental sustainability and create conditions of social marginalization'. But it is not enough to simply include citizens in rebuilding processes. The issue of trust is paramount in any effort toward constructive consensual post-disaster efforts and those relationships are more easily developed in pre-disaster planning phases. Of course, according to Allen and Maret (this volume), '[t]rust is a reciprocal exchange and unfortunately in the case of many governmental agencies in post-Katrina New Orleans was institutionalized as distrust in its citizens'. Trust can be gained in post-disaster scenarios by outside aid organizations, but concerted efforts need to be made to achieve a level of goodwill and openness immediately upon engaging with the displaced and repatriating residents. Successful strategies used by many NGOs in post-Katrina New Orleans included allying themselves in a meaningful way with trusted local community groups that had worked positively within the city before the storm, and expanding their existing capacity to include post-disaster rebuilding programmes.

Implications for the dynamics of disaster

Many emergency managers and planners tend to revolve disaster planning and mitigation around available technologies with the rationale that cutting edge technologies will help to achieve better preparedness and response. It is around such tendencies that the intersections of Science and Technology Studies (STS) and disaster science and management need to be reinforced. Technologies are only as effective as the social and political networks that use them for risk assessment, planning and responding to disasters. Practitioners in emergency management are only too familiar with failed plans due to communication breakdowns that had little to do with technological failures, but more due to institutional barriers to the flow of information. It is therefore important that practitioners, disaster researchers and STS researchers ground themselves in lived experiences when confronted with standardized plans that do not translate well to different locations, communities and/or times. Questions about the politics of scientific and technical knowledge as well as questions about their intersections with social and cultural domains of lived environments are foregrounded in STS analytical lenses. This volume has shown that cultures of agencies, politics of institutions, social networks and situated knowledge and history shape post-disaster outcomes in significant ways. However, the application of these lessons is only as effective as the networks among practitioners and researchers. Lived experiences are the ties that bind these groups, as much as they are what binds networks that determine the dynamics of disaster.

Index

Page numbers in *italics* refer to figures, tables and boxes